Sereno Edwards Todd

The Apple Culturist

A Complete Treatise for the Practical Pomologist

Sereno Edwards Todd

The Apple Culturist
A Complete Treatise for the Practical Pomologist

ISBN/EAN: 9783337399245

Printed in Europe, USA, Canada, Australia, Japan

Cover: Foto ©berggeist007 / pixelio.de

More available books at **www.hansebooks.com**

THE

APPLE CULTURIST.

A COMPLETE TREATISE

FOR THE PRACTICAL POMOLOGIST.

TO AID IN PROPAGATING THE APPLE, AND CULTIVATING AND MANAGING ORCHARDS.

ILLUSTRATED WITH ENGRAVINGS OF FRUIT, YOUNG AND OLD TREES, AND MECHANICAL DEVICES EMPLOYED IN CONNECTION WITH ORCHARDS AND THE MANAGEMENT OF APPLES.

By SERENO EDWARDS TODD,

AUTHOR OF "TODD'S YOUNG FARMER'S MANUAL," "AMERICAN WHEAT CULTURIST," "TODD'S COUNTRY HOMES," AND "HOW TO SAVE MONEY."

NEW YORK:

HARPER & BROTHERS, PUBLISHERS,

FRANKLIN SQUARE.

1871.

DEDICATED

TO THE

REV. THEODORE LEDYARD CUYLER, D.D.

MY HIGHLY - ESTEEMED FRIEND,—Understanding, in a measure, your cheerful zeal and co-operation in every philanthropic enterprise which tends in any way to render the world wiser, and mankind better and happier, and knowing your appreciative taste for luscious apples, permit me to dedicate to you a little volume on the culture of your favorite fruit.

The foundation of our holy religion lies in the virtuous industry of the people. Hence, in our efforts to render the human family the recipients of the greatest good, we need to educate their lower faculties *first*. If we teach a nation to culti-vate bountiful crops of fine wheat, and to produce large supplies of excellent apples and other fruit, we have secured a foundation on which it will be compara-tively easy to develop the finer and nobler faculties of the soul. Taking this view of the duties of our mortal state, it is a cheering thought that, while we played and ate apples together in the days of boyhood, we may labor side by side—I at laying the foundation, and yourself in lifting up and fortifying a glori-ous superstructure of manhood—in society, where virtue, religion, and truth are the crowning excellences.

If we go where no boughs laden with choice fruit bend to kiss the rosy cheeks of guileless children playing beneath, and where no waving grain rolls in the summer breezes like a sea of gold, we shall find a pall of heathenish darkness resting on the people like a mighty incubus. Hence, I send out this little book to the world, with the hope that it will perform the duties of a philanthropic pio-neer in preparing the rough ways of civilized life for the more complete enjoy-ment of an elevated manhood. I trust it may be found a timely *vade mecum* in the hands of young men who have a desire to establish happy homes and to cul-tivate choice fruit.

Superb apples are the product of Eden. A good boy with a hatful of Sweet Boughs and a pocketful of gingerbread will always be found a more tractable pu-pil when getting his lessons in the Catechism, than if the stomach were distend-ed with heavy animal food of a stimulating character. A home without children and destitute of apples, is like a beautiful grove without the cheering songs of birds. What delightful memories dance in the sunshine of our boyhood, as our thoughts revert to the homes of our early years, on fair Cayuga's fertile slope in Central New York, where the pathway of life was embellished with apple-trees which seldom failed to shower down golden luxuries in great profusion! Those were halcyon days in our happy experience. Fond memory delights to linger in the extensive apple-orchards where bountiful supplies of Sweet Boughs, Swaars, Spitzenbergs, and other choice varieties rendered material aid in smoothing the asperities of our buoyant existence; and we often wish we were boys again— if it were possible to begin a new career with our present experience—that we might again rejoice in the delight which once swelled the young heart at the sight of hatfuls and pocketfuls of ruby apples.

With my best wishes for your success in all your labors of love, and that your last days may be replete with joy and gladness,

I remain your faithful friend,

SERENO EDWARDS TODD.

Brooklyn, L. I.

SERENO EDWARDS TODD, Esq.:

MY DEAR OLD FRIEND AND SCHOOL-MATE,—I thank you heartily for the pleasant compliment of linking my name with your savory treatise on my favorite fruit. The very reading over of that goodly catalogue of varieties—from the "Early Harvest" and the "Strawberry," on to the "Newtown Pippin" and the "King-apple" —carried me back to the cellar and the apple-bin of my boyhood. When you and I went to the district-school together, we crammed our pockets with "Swaars" or "Greenings" for the noonday lunch. What French confections are to city-bred children, that were a hatful of apples and a pocketful of hickory-nuts to us homespun lads in the dear old free, broad country. A book that recalls those days is as "sweet to me as the breath of new-mown hay."

May your latest volume be as popular and useful as its many predecessors from your fertile pen, wisheth

Your friend of yore,

THEO. L. CUYLER.

Lafayette Avenue Church, Brooklyn, Dec. 15, 1870.

PREFACE.

I sing of the apple, with roseate bloom,
That flourished in mazes of verdurous gloom
In Eden's fair bowers—in tint and in shape—
The apple that vies with the peach and the grape.—EDWARDS.

FROM early boyhood the writer of this little treatise has been *practically* engaged, more or less every year, in the propagation of apple-trees and the management of orchards. As there is no little manual on "Apple Orchards" in all our agricultural and pomological literature which a beginner may study as a reliable guide in every branch of apple culture, and as the author in early life was obliged to advance from one step to another, in the rearing and management of apple-trees, by the slow and often uncertain way of determining the better practices by experiments, he feels warranted in preparing a small volume, in which are embodied the successful results of the experience of more than forty years.

Most of the works on pomology are either too voluminous and expensive to meet all the requirements of men who need a cheap book containing brief directions for beginners, or the writers have *assumed* that their readers *already* possess a pretty correct understanding of that branch of pomology which treats of the correct mode of propagating apple-trees, and the most satisfactory manner of managing apple orchards in order to produce profitable crops of fruit.

Every beginner will always encounter certain difficulties in the production and management of apple orchards ; and he must necessarily learn how to overcome them in a reliable and satisfactory manner, either by experiments conducted frequently, amidst perplexing doubts as to the ultimate results, or he must be furnished with the results of numerous experiments made by practical pomologists, which he can rely upon with the same confidence that he would feel if he had worked out the same results on his own grounds. The aim of the author, therefore, has been to supply inquiring beginners with the fundamental knowledge which one *must* possess before he can proceed satisfactorily in the cultivation of apples, and to record such results of long experience as will enable any intelligent person to perform the operations required in the management of orchards, or of the apples, in the best and most approved manner. Hence a satisfactory answer to almost any question that an inquirer after pomological truth touching the apple may desire to have practically elucidated, may be found in some one of the chapters of this little book. Reliable facts, conveyed in plain and intelligible language, have moved the author's pen, rather than any desire to roll out beautifully-rounded sentences to please the fancy more than they would instruct an humble inquirer after truth.

There are many thousands of young men in all parts of the country who need the aid that such a practical treatise on the apple will furnish. The writer has endeavored to present every subject in such a manner that a beginner will be able to perceive and to appreciate what should be done, as well as what is not allowable. Apple orchards seem to be failing—for which there are plausible reasons ; and we have endeavored to show intelligent beginners the *true causes* of failure, to direct their operations in such a manner that there shall be no such thing as a failure of the apple crop. We have recorded nothing that has not been put to a practical test, and found, by long experience, to be reliable. We have also endeavored to encourage beginners to plant an orchard early in life, and to manage the trees in such a manner that they will never lack a liberal supply of good apples.

The author has been writing more or less on fruit for thirty years past ; and some of the articles have been published in the " Cultivator " and " Country Gentleman," in the " American Agriculturist," " New York Times," and " New York Observer," while editorially connected with those journals ; some in " Moore's Rural New Yorker " and in the " Working Farmer," all of which have been reconstructed and revised. With the exception of a few small illustrations reproduced from electrotypes taken from " Downing's Fruits and Fruit-trees of America," and a few also from the " American Entomologist," through the courtesy of its publishers, the illustrations have all been engraved by the publishers of this book.

SERENO EDWARDS TODD.

Brooklyn, L. I.

CONTENTS.

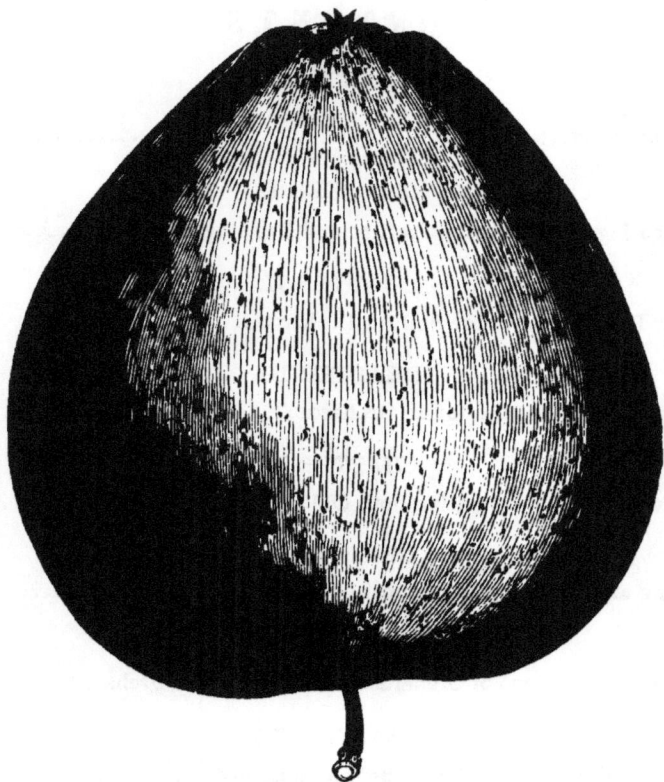

YELLOW BELLFLOWER.

Synonyms.—Belle Fleur, Bellefleur Yellow, Lady Washington, Warren Pippin, Bishop's Pippin of Nova Scotia, and Reinette Musque. The engraving, as to form, is a fair representation of the Yellow Bellflower. The fruit is usually large and beautiful; skin smooth, pale lemon yellow, often with a beautiful blush on the side next the sun. The flesh is tender, juicy, crisp, with a slightly sub-acid flavor. Before the fruit is fully ripe, it is considerably acid. Season, November to March. The trees are excellent bearers, and moderately hardy. This variety commands a ready price in market, as the fruit is large, and usually handsome. The Yellow Bellflower is an excellent apple for pies and apple-dumplings.

INTRODUCTION.

We come with a handful of apple-seeds red,
To plant with the spade in a deep mellow bed:
When tempests of winter, in fury sweep by,
On earth's gelid bosom the tiny seeds lie.—EDWARDS.

APPLES of some varieties are cultivated successfully over a greater breadth of country than any other fruit. No other trees can be relied upon for a regular supply of choice fruit every year, for a great number of years, with more certainty of a crop, than apple-trees. On every hill and in every valley, on every plain and mountain between the cold and backward localities of Maine and Canada, to the extreme border of the Golden State, apple-trees, if correctly propagated and properly managed, will not fail to reward the tiller of the soil with abundant crops. More than this, there is no other kind of fruit cultivated that can be made to mature during such a long succession of months as the apple. And then the almost unlimited variety as to the taste and quality of fruit, constitutes another consideration of transcendent importance. We have varieties which are fully ripe and delicious before the last days of rosy summer have faded away, and a regular succession afterwards of choice varieties that are in season until there is an encouraging promise of another crop. Apples are one of the greatest luxuries that people can depend upon as a regular article of food. They are excellent while in a crude state, and superb when cooked in a score of different ways. Adults like them, and children love them. All kinds of

1*

domestic animals, from the noble carriage-horse down to
the cackling poultry, will devour apples—especially sweet
ones—with great avidity. And such food, in connection
with other articles, will always be found profitable.

> "For dumplings, pies, or even apple-stew,
> What could our cooks or our good housewives do,
> If by perchance the apple should step out?
> Sore grief would seize on many bosoms stout."

An eminent French physician says the decrease of dys-
pepsia and bilious affections in Paris is owing to the increased consumption of apples, which fruit, he maintains,
is an admirable prophylactic and tonic, as well as very
nourishing, and easily digested.

The proper contemplation of the production of a large
apple-tree, from a tiny seed, with branches bending beneath
a bountiful crop of luscious fruit, is a source of sublime
and interesting thought to every reflecting mind. The
swelling of the germ in the seed-bed, the bursting of the
envelope of the kernel, the starting of the thread-like roots,
and the formation of a stem, constitute *germination* of the
seed; and the more complete growth and development of
the radicles, stem, branches, leaves, and flowers embrace the
idea of *vegetation*. The unfolding of the fruit-buds, the
development of the flowers, and the perfection of the fruit,
constitute *fructification*.

All these operations, when taken collectively and studied,
so that the different processes of growth and development
shall be appreciated, will give the intelligent pomologist
such an insight into fruit-growing as will tend to supply
our markets, and the tables of both rich and poor, with an
abundance of the cheapest luxuries of mortal life. The
production of a fruitful apple orchard, from the time it is
set until it comes into bearing, is necessarily a question of
time—to say the least, requiring from five to eight years,

under the most favorable circumstances, to obtain any thing like a supply of fruit for the use of a family.

There is no way a man can so effectually rear a monument to posterity, and one for which he shall receive their blessing, as by planting a fruitful apple orchard that shall yield a luxury in its wealth of delicious golden fruit. To the young, there is nothing about the farm more attractive than the orchard. They have a constant, insatiable appetite for fruit — for apples, peaches, cherries, pears, nectarines, grapes, melons. The country is not the country without them. The green fields, the majestic forest-trees, the cool springs and the meandering streams, are all engraved indelibly upon their memories. And when they review the happy days of childhood, and think of

"The *orchard*, the meadow, the deep-tangled wild-wood,"

fond memory lingers longest about the old apple-trees that have so often filled the pockets, hat, and baskets with bountiful supplies of luscious fruit.

The production of a fruitful apple orchard, and the annual care and protection of the growing crop of fruit, constitute not only a delightful, elegant, and scientific occupation, but, where the locality is such as to afford market facilities, it is really one of the most profitable in the whole range of agriculture. Good ripe apples are not only objects of great beauty, highly conducive to health, but they are an important article of domestic consumption and household economy. They furnish the best and cheapest desserts, that enter largely into a variety of dishes. They supply the table during the summer and fall months, when crude, dried, and preserved, and they last through the winter. Good apples are at all seasons, and in all their various forms, useful and delightful, both in health and in sickness. When we contemplate the perfection to which apples have

been brought by proper care and culture, we are amazed at
the neglect, the utter and almost criminal neglect and con-
tempt of which they have, for two centuries, been the subject in the states of the South, and to a great extent now
are.   The idea has prevailed, and still prevails, that we
have a claim upon nature for a fruit crop without assisting
in its production; and when it fails we blame nature, the
trees, the soil—any thing and every thing but ourselves, to
whom the blame chiefly belongs.   Every man who has a
farm or a garden should raise fruit, and attend carefully to
his trees.   If he raises it only for domestic use, he will be
repaid for his trouble tenfold.   If so situated as to be able
to throw it into market, he will find it one of the most
profitable of his crops.   Under the old plan of planting an
orchard and leaving it to itself, a generation passed before
it was productive; but under modern culture it yields its
fruits in a few years.   For instance, we find it stated by a
reliable authority, that in a single garden, apple-trees, the
fifth year from setting out, yielded a bushel each; peach-
trees, the third summer, bore three pecks; and a Bartlett
pear, two years from transplanting, gave a peck of superb
fruit.   None of these trees were an inch in diameter when
transplanted, nor was their treatment better than that
which every good farmer bestows upon his carrots and
potatoes.   The apple orchards of America are a striking
characteristic of the energy, ambition, and persistent utility
of a nation which is a unit in power, efficiency, and nobility,
which can not be found on any other portion of our globe.

Hon. Horace Greeley once wrote: "If I were asked to
say what single aspect of our economic condition most
strikingly and favorably distinguished the people of our
Northern States from those of most if not all other countries which I have traversed, I would point at once to the
fruit-trees which so generally diversify every little as well

as larger farm throughout these States, and are quite commonly found even on the petty holdings of the poorer me-
chanics and workmen in every village, and in the suburbs
and outskirts of every city. I can recall nothing like it
abroad, save in two or three of the least mountainous and
most fertile districts of Northern Switzerland. Italy has
some approach to it in the venerable olive-trees which sur-
round or flank many, perhaps most, of her farm-houses, up-
holding grape-vines as ancient and nearly as large as themselves; but the average New England or Middle State
homestead, with its ample apple orchard and its cluster of
pear, cherry, and plum trees surrounding its house, and dot-
ting or belting its garden, has an air of comfort and mod-
est thrift which I have nowhere else seen fairly equalled.

On the whole, I deem it a misfortune that our Northern
States were so admirably adapted to the apple and kindred
fruit-trees, that our pioneer forefathers had little more to
do than bury the seeds in the ground and wait a few years
for the resulting fruit. The soil, formed of decayed trees
and their foliage, thickly covered with the ashes of the prim-
itive forest, was as genial as soil could be; while the remaining woods, which still covered seven-eighths of the
country, shut out or softened the cold winds of winter and
spring, rendering it less difficult, a century ago, to grow
fine peaches in Southern New Hampshire than it now is
in Southern New York. Snows fell more heavily, and lay
longer, then than now, protecting the roots from heavy
frosts, and keeping back buds and blossoms in spring, to
the signal advantage of the husbandman. I estimate that
my apple-trees would bear at least one-third more fruit if I
could retard their blossoming a fortnight, so as to avoid
the cold rains and cutting winds, often succeeded by frosts,
which are apt to pay their unwelcome farewell visits just
when my trees are in bloom, or when the fruit is forming

directly thereafter. An apple orchard in full bearing, the tempting fruit blushing among the foliage, or covering the ground with a profusion of golden "nuggets" such as no mine can yield, has a fullness of beauty which, while it charms the eye, appeals not less successfully to other senses.

Hon. H. T. Brooks once said in an agricultural address, when alluding to the value of the apple: "By no earthly process, in my opinion, can so much nutriment be so cheaply extracted from four square rods of ground as by planting an apple-tree in the centre, and giving it good cultivation. Apples need the ground, the whole of it, and all it contains, but "immemorial usage" allows an apple-tree no rights that husbandmen are bound to respect. It is haggled and mangled, roots and branches, and the soil exhausted in the production of other crops. Charging the apples with the ground they actually grow upon and appropriate, they give far better returns as food for man or beast than corn, wheat, or potatoes. New York, particularly Western New York, has a character at home and abroad for fruit. If a better apple country was ever made, I confess I never heard of it. We occupy the precise position where the tree is hardy and healthy, and the fruit comes nearest perfection. I know of no ordinary farm crop that at all compares, during a series of years, with apples, if we take into the account the small expense at which they are raised. Should we reduce the yield to one half-barrel to the tree, apples would still be our most profitable crop. I boldly claim that the average of our orchards could be doubled by good cultivation. An acre of ground that will produce forty barrels of good fruit ought to be excused from growing grain. Whatever grain or root crops are grown upon it, detract, doubtless, more than they are worth from the apple crops. We can not, without great expense and trouble, return to the soil all the elements which our wheat, corn, and potatoes take from it.

When I hear of trees standing near a wood-pile, in the corner of a fence, near the barn, or the hog-pen, or the kitchen-door, I am prepared for a big yield. The great majority of our apple-trees are either starved or go very hungry. A free application of barn-yard manure is indispensable to the continued growth and productiveness of our orchards. Ashes, lime, plaster, and perhaps other mineral fertilizers, may be used to great advantage. It should be remembered that the apple-tree is subject to the general laws of vegetable growth. You can not have large and fair fruit if the soil is poor, hard, and dry. The earth should be mellow, to secure suitable moisture in hot, dry weather, and to impart needed nourishment. Begging pardon of the Pomological Society, a tree knows better where to put its roots than any man can tell it; and the apparently stupid roots understand far better than any of our reputedly wise chemists what elements are essential for the production of a bountiful crop of fine fruit."

A Succession of choice Apples.—Every family that is in possèssion of only a few roods of good ground—even if much of the surface be rocky and comparatively untillable —should have a succession of crude apples, suited to every season of the year, adapted to different tastes and to various household uses. If a person has the land, there can be no possible excuse for not having a bountiful supply of superior fruit in six to ten years, unless we accept the shallow pretext so often advanced, "a want of time" to cultivate the trees. Every man fritters away every season, to no satisfactory purpose, far more time than would be required to plant an orchard and take proper care of the number of trees requisite to supply his family with crude apples during the entire year. The apple is quite different from perishable pears, peaches, plums, and other kinds of fruit, which must be consumed to-day, or they will be worthless to-morrow. The choice varieties of apples are now so extensive that, by proper management, in our latitude—New York city—any family that will appropriate only a part of one acre to a few trees of good varieties which will mature in succession may begin to gather crude, ripe apples about the first of July, while they may still have in the cellar a small supply of *old* apples. When on the farm, we frequently ate new apples of the Early Harvest variety and Roxbury Russets on the same day, even when we had no facilities for keeping apples, except a good cellar beneath the dwelling. We give herewith the names of a few varieties which will furnish a succession from the middle of July of one season, to the same period — or even later— of the following season : Early Harvest, Tallman Sweeting, Early-Sweet Bough, Fall Orange, Early Chandler, Williams, Garden Royal, Porter, Gravenstein, Mother, Hubbardston's Nonsuch, Rhode Island Greening, Ladies' Sweet, Peck's Pleasant, Baldwin, Roxbury Russet, Early Joe, American Summer Pearmain, Benoni, Early Strawberry, Red Astrachan, Summer Pippin, Duchess of Oldenburgh, Twenty-Ounce, Hawley, Tompkins County King.

One tree of each of the foregoing varieties, if properly cultivated, would supply a small family with all the fruit they would need during the year, before the trees are half-grown. Those persons who desire extensive orchards can add other varieties to suit locality or the market. By referring to the voluminous treatise, "Downing's Fruit and Fruit-trees of America," the reader will find a description of almost every known variety of apples, a list of which it is impracticable to give in this small work.

The man who desires to have a good apple-orchard has only to avail himself of the facilities within his reach on any soil between the Atlantic and the Pacific Oceans, where it is practicable to raise fair crops of cereal grain ; and apples in copious abundance can be produced on dwarf trees in four or five years, and on standards in six to ten years. A well-balanced brain and a skillful hand will not fail to produce fine fruit.

We have in mind an old farmer in Ohio who felt prompted, when a young man, to plant an orchard; but he had imbibed the erroneous notion that the man who plants apple-trees seldom lives to partake of the fruit. When at the age of forty, fifty, sixty, and at seventy, he looked back with regret that he did not plant an orchard when he was a young man. As he passed the age of threescore-and-ten, he resolved to plant an orchard; and the trees came into full bearing so soon that he lived to eat the luscious fruit of his labor for several years, and to get drunk on the cider made from the apples of those trees which were planted at such a late period in his life. Apple-trees, like our children, will grow up so quickly that we are surprised to contemplate how soon they are filled with fruit. If one-half the money that is now expended by the laboring classes for tobacco and intoxicating beverages, the pernicious influences of which fill the land with crime, and spread unhappiness and desolation around the fireside, were employed to cultivate apple-trees, or to purchase fruit, many apothecary shops would be closed at once for lack of patronage; quacks and doctors would be obliged to seek other employment; and unhappy homes would be changed to places of delight. A small fruit-orchard is far more valuable, for any family, than bonds, mortgages, or money at legal interest.

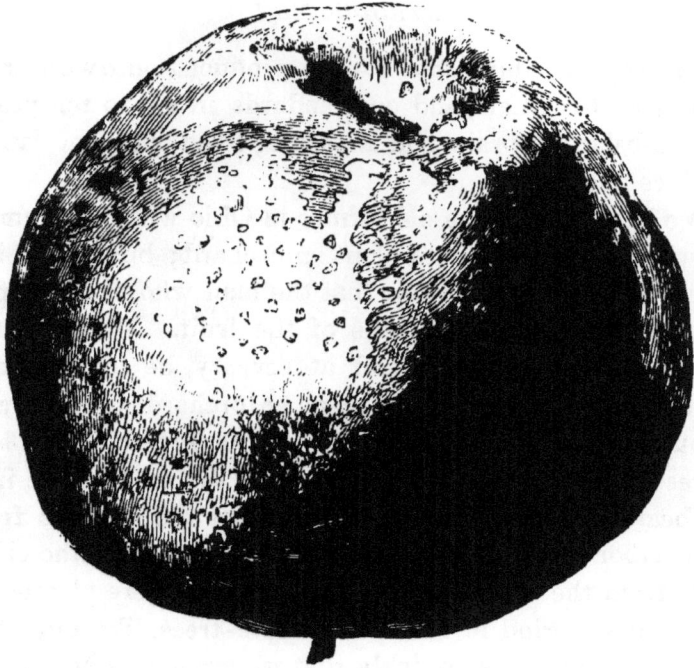

RED CANADA.

Synonyms.—Poland, Richfield Nonsuch, Steele's Red Winter, Old Nonsuch of Massachusetts. This is an old variety, which has been almost run out by injudicious management of the orchards. With good cultivation the tree is very productive. The fruit is of medium size, oblate, inclining to a conical form. Skin yellow, and usually shaded with deep red crimson, somewhat striped or splashed on the sunny side. The core is small and close. Flesh white, tender, crisp, abounding with a refreshing juice. In quality, superior. Season, January to May.

THE

APPLE CULTURIST.

CHAPTER I

PROPAGATING APPLE-TREES.

"Here embryo apples in tough rinds compressed,
Are folded in beauty, on each floret's breast.
The sunshine of April and breath of sweet May,
Shall lift up the plumule and spread out the spray."

THE apple is propagated by planting the seeds, by grafting, and by inoculating, or budding; and it *may* be propagated by *cuttings* and *layers*. The core of an apple is frequently thrown aside, where a seed sometimes finds a mellow seed-bed; and the next season a young tree appears. Birds frequently drop apple-seeds in a bed of fine mould, where the embryo is preserved, until circumstances favor germination. Young apple-trees frequently spring from the seed in the remains of decayed apples.

Apple and pear seeds are usually collected in autumn, separated from the pomace, mingled with moist sand, and kept in a cellar till the following spring, when they are planted in drills as one plants beet-seeds, in a well-prepared seed-bed. In some instances, the pomace is scattered in drills, and covered with fine and mellow soil, in late autumn. When large quantities of young trees are required, the seeds are sometimes sown broadcast like grain, and harrowed in; and in many instances the seeds are put in with grain drills, about as deep as seed-wheat and barley are covered—say two inches deep in a mellow seed-bed. We have in mind a friend in Illinois, who wrote that he would drill in forty bushels of apple-seeds in the spring of 1870, distributing that quantity over eighteen acres.

As there is a tough covering on the outside of apple and pear seeds, they will germinate more readily if they can be planted in autumn, or early in the spring, so that they may be frozen and thawed two or three times before the growing season has commenced. And yet, apple-seeds will germinate without having been frozen, provided they are not allowed to become dry after being separated from the core. If seeds be planted as soon as they are taken from the core, they will often germinate, and appear in the seed-leaf in eight to twenty days, according to the moisture and warmth of the seed-bed. We have taken seeds from an apple, planted them in a mellow-bed, in June, some of which germinated immediately, and others remained in the ground till the following spring. Some pomologists contend that apple-seeds, and seeds of other fruits, should be buried in *the fruit*. But as there is usually a chance to select the best kernels if the cores are first dissected, which is quite as important as to choose the most desirable ears of Indian corn for seed, when this is done, the fruit should be cut in quarters, the inferior kernels separated, one or two of the largest and plumpest returned to the core, and the entire apple buried about two inches deep in a mellow seed-bed.

Raising Seedlings.—The practice of raising seedlings, or young stocks, from the seeds by planting the half-developed kernels of the half-grown and half-matured fruit of a miserable variety of the poorest kinds of cider-apples, which swine would almost refuse to eat, is about like attempting to produce fine cattle from the meanest and most miserable scrub that could be found in the country. Like will pro-

duce like. The stock of a grafted tree will exert a wonderful influence on the productiveness of the bearing tree.
(See *Glossary.*) It can not be expected that the half-developed seed of a half-grown and half-ripened, small, knotty, scabby, one-sided, worthless apple can ever produce a
fine and prolific tree. Is it possible for apple-seeds to impart certain prominent characteristics of excellence to the
future tree and fruit, which were never possessed by that
variety? The truth is, that the best seeds must be selected for fruit-trees, and the inferior kernels rejected, if we
would produce hardy and prolific trees.

How Apple-seeds vegetate. — An apple-seed germinates
like the kernels of leguminous plants, such as beans, clover, and flax. The embryo (Fig. 1) expands,
and a radicle appears at the pointed end of the
seed, which grows downward, as the stem is
formed upward. Instead of a spike being sent
to the surface of the ground, such as appears
from a kernel of cereal grain, the apple-seed is
thrust upward, on the end of the stem, through
the soil, to the surface of the ground. This
fact suggests the eminent importance of covering every seed with fine mould or mellow earth.

Fig. 1.

Enlarged view
of the inside
of an apple-
seed.

A covering of heavy clay will often become so compact
that the stem can not elevate the kernel to the surface. As
soon as the apple-seed appears above the surface of the
seed-bed, the kernel separates into two equal parts, each
portion being held by the stem; and the two lobes assume
the form of leaves. The accompanying illustration (Fig. 2,
p. 22) is a fair representation of a young apple-tree soon
after the kernel has appeared above the surface.

There are several fundamental requirements to be observed by the tiller of the soil, in order to effect the germination of apple-seeds. In the first place, a certain amount

Fig. 2.

A young apple-tree in the seed-leaf.

of moisture must reach the germ. If the external coating of the seed has been allowed to become very dry and hard, seeds will often remain in the soil, if planted in the spring, until the next spring, before the tough shell will permit moisture to quicken the germ. When apple, locust, and other seeds have become dry, let them be thrown for ten seconds or so into boiling water, then turned quickly into cold water, repeating the process twice or thrice, until the hard shell is softened, and the seed will germinate in a few days after they are planted. By a short and quick scald, and sudden cooling, the heat does not have time to reach the germ. But let me caution every beginner against the mistake that a celebrated agricultural editor once made in boiling his seeds for several hours, to soften the hard shells. The shells of chestnuts, butternuts, and other nuts, as well as peach and plum pits, are so impervious to water, that moisture can not reach the germs. Hence the necessity of soaking a long time, and exposing these to the action of frost, in damp mould, or sand, so that the glue which unites the two parts of a shell may be dissolved, thus permitting the moisture to reach the embryo or germs.

Fig. 3.

A diminished view of a young apple-tree after a few inches in length of the plumule have appeared.

Another indispensable requisite is *heat* in connection with moisture. Gentle warmth alone can not produce germination; neither can moisture alone cause a seed to germinate. The two—moisture and heat—must exist in a proper de-

gree, or apple-seeds will never germinate. Still another condition essential to germination is *warm air*. Apple-seeds, if buried too deep, where they are supplied with moisture and heat, *without air*, will soon mould and decay. Three things, then, must all combine harmoniously to promote germination, or an apple-tree can never be produced from a seed.

Propagation by Grafting.

But when the smoother stem from knots is free,
We make a deep incision in the tree;
And in the solid wood the slip inclose,
Where it unites and shoots again, and grows.—VIRGIL.

Grafting is the insertion of a cion in a living stock. The philosophy of grafting consists in making a cleft, or slit, in the end or side of the stock, and fitting one end of the cion to the cleft so neatly that the pores between the bark and the wood of the cion will correspond with the similar pores in the stock. When cions are inserted in a stock in this manner, they can scarcely fail to grow. It is of little consequence how a cion is grafted, provided the inside bark of both cion and stock coincide, so that the flowing sap may readily pass from the stock into the cion. But the cion and stock must be united with such precision that the surface of one will fit the surface of the other, water-tight. When a bad fit is made, the surface will soon oxidize, and prevent all union.

Modes of Grafting.—The different modes of grafting are alluded to as *cleft* grafting, *whip* grafting, *American whip-tongue* grafting, *splice* grafting, *shoulder* or *chink* grafting, *crown* grafting, *saddle* grafting, *side* grafting, *dovetail-side* grafting, *summer* grafting, *root* grafting, *stock* grafting, *spur* grafting, and grafting by *approach*, or *inarching*, which is a curious way of attaching a portion of the cion to the stock to which it is to be united, while another portion still remains on the parent stem. Branches of two different trees may be united at the extremities, "by approach," like Fig. 23, if the parts be neatly fitted, and held firmly by means of splints, until the union is complete. The different modes will be shown by illustrations.

Grafting into different Species.—Beginners who do not possess a correct understanding of the laws of both vegetable and animal physiology are often ambitious to try experiments in grafting or budding a given kind of fruit on a tree of some other species. In numerous instances, by not understanding what *can* be done, and what is impracticable, they have committed ludicrous blunders. As stated under *Species*, apples may be produced by inserting apple-tree cions, or buds, in a pear-stock. And pears, in the same manner, may be produced on apple-stocks. We have often seen pear-cions inserted in "thorn-apple" stocks; and have frequently read recommendations, by writers on pomology, as to grafting pears on thorn-stocks. But we have never met with, nor read of, satisfactory success in such experiments. Pears and quinces, also, belong to different *species* of fruit; and yet the product of pear-cions on a quince-stock is a satisfactory success. Apples, pears, and quinces have so little affinity with peaches and plums, that they can not be produced on a peach or plum stock. Nor can peaches or plums be produced on apple-stocks.

Apple-cions may be set in the maple or willow; but were they to live and grow the branches would never yield fruit. We have frequently seen it stated, in agricultural journals, that apples and pears have been produced satisfactorily on the young stocks of the mountain ash; but we have never met with a person who has seen the fruit. A gentleman near Goshen, N. Y., assured us that he had been successful in grafting the cions of the English walnut on his young sweet-walnut-trees. But walnuts will not grow on the chestnut or butternut trees, nor, *vice versa;*

simply because there is not sufficient affinity between the stocks and the cions.

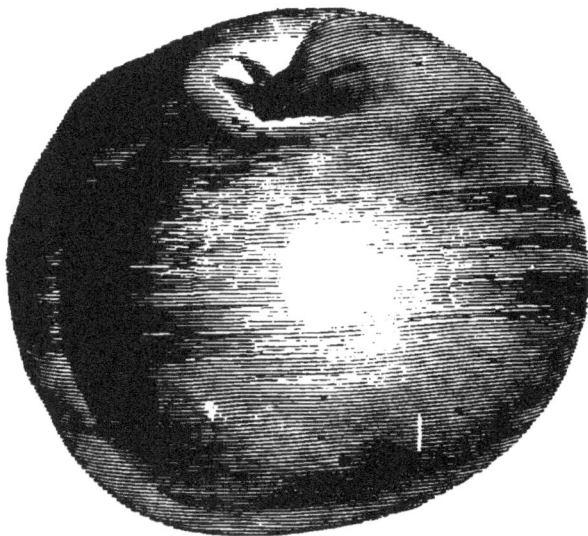

FAMEUSE.

Synonyms.—Pomme de Heige, Sanguineus, Snow. *Fruit.*—Size, medium; form, roundish, somewhat flattened; skin, smooth; color, a greenish-yellow ground, mostly overspread in the sun with a clean, rich red; in the shade the red is pale, streaked, and blotched with the dark red; stem, slender; cavity, narrow and funnel-shaped; calyx, small; basin, narrow and shallow; flesh, remarkably white, tender, juicy, negative character, but deliciously pleasant, with a slight perfume; core, close, small, compact; seeds, light brown, long and pointed; season, October, and to December. *Tree.*—Hardy, healthy, moderate grower, of a rather diverging habit, with dark-colored shoots, and long, narrow leaves, bearing annually a fair crop, with a profusion in alternate years.

Propagating by Root-grafting.—Volumes have been written touching this subject, to show that root-grafted trees will not endure so long as other trees which have sprung from grafted stocks. In many of our Western States, reports have been made by practical pomologists, who have instituted inquiries and numerous experiments to test the duration of such trees; and in most instances that have come under our observation, root-grafted orchards have nearly all failed after a few years, especially on large prairies. In some instances, there has been no apparent difference, which may be accounted for by the fact that in these cases strong tap-roots were sent down deep into the sub-soil, which supplied the growing trees with sufficient moist-ure. It will doubtless be found that, when root-grafted trees fail, they are destitute of tap-roots, or large-branch-ing roots, to supply moisture in dry weather, when the roots near the surface of the ground can not obtain the necessary amount.

Any one at all conversant with the habit of the different varieties of apple-trees knows that there is a great differ-ence in the growth and hardihood of the original stock. This subject is not half enough considered by the orchard-ist. Let him go into a nursery where all the different vari-eties are growing in a good soil, each shading and protect-ing the other thickly in the rows, under good, and frequent-ly forced cultivation, and he will at once suppose they are all alike thrifty, hardy, and promising. But such is only the fact while in the nursery. Some varieties are tender, slow of growth, and never hardy when exposed in the orchard to the fierce heats and cold blasts which alternate-ly shine upon and sweep over them. Other varieties are hardy, vigorous, and stalwart under all circumstances. Others still there are which bend and writhe about, scarce-ly knowing which way to grow. These different modes of growth are original properties of the wood itself, nat-ural, organic, and only to be corrected or overcome by care and attention in the subsequent training of the trees.

An intelligent pomologist writes that the result of this process of raising trees by root-grafting is, that after ten or twelve years standing in the orchard, with equal care and cultivation, some trees are twice or thrice the size of others. Some are feeble and decaying, from innate weak-ness or exposure to outside influences; while others are strong and vigorous as oaks or maples. We think it will

be proved, *as a rule,* that fruits of high quality are usually more refined and delicate in their wood than those of
coarser and harsher taste; that the common seedling is usually hardier in its stock than the highly cultivated "graft;"
and, therefore, that the common seedling reared up to a
size fit for transplanting into the orchard, and then grafted
branch high, or at the point where its limbs diverge into
the branching top, is better as stock than those which are
root-grafted.

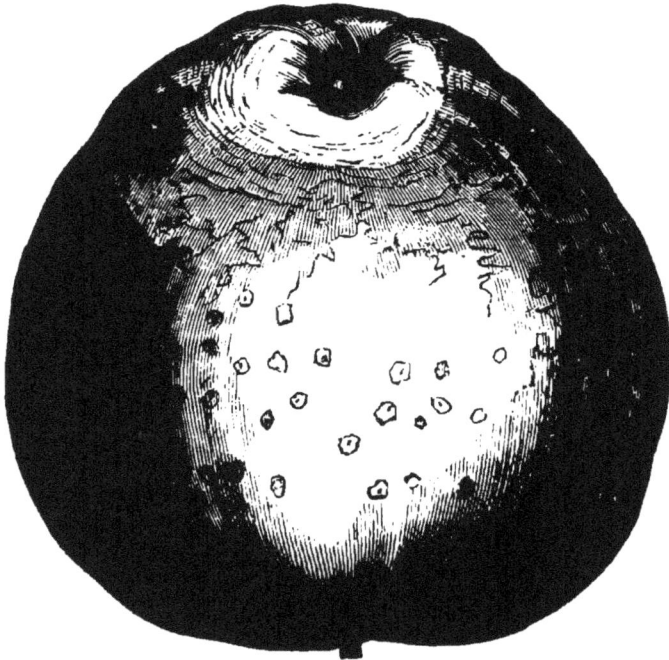

WESTFIELD SEEK-NO-FURTHER.

Synonyms.—Seek-no-further, Red Winter Pearmain, Connecticut Seek-no-further. *Fruit.*—Size, medium; form, regular roundish conical, broad at base; color,
generally a light yellow ground, with the sunny sides striped and splashed with
red; small russet dots, surrounded with shades of a light russet-yellow; there is
often considerable russet around both stem and calyx; stem, long and slender;
cavity, open, regular; calyx, usually small, and generally closed, or nearly so; it
is, however, sometimes partially open, and always with short segments; basin,
regular in form, and of moderate depth; flesh, yellowish, tender, sub-acid, with a
pleasant Pearmain aroma; core, medium; seeds, ovate. Season, November to
March.

A hundred years ago, pomologists seldom thought of propagating apple-trees by root-grafting. When they ate a good apple, their first idea was to plant the seed. By this practice, and by not cutting off the tap-roots, they obtained trees that were bountifully productive for a long time. The bare fact that pomologists have had occasion to feel assured that root-grafted trees are sometimes *not* so productive as seedling stocks, should be sufficient reason for abandoning entirely that manner of propagation.

This question has been discussed extensively in Illinois; and many intelligent pomologists would not accept root-grafted trees as a gratuity. One gentleman stated at the pomological convention that he had made observations for several years, and had come to the conclusion that, for *that* region, trees worked standard high are better worth a dollar a tree than such as are root-grafted with a dollar as a gratuity. He had found such shy bearers as Early Harvest, Pryor's Red, etc., to bear well when worked on stocks. The rule had been found general, with some exceptions. He had found budded trees to bear good crops in six to eight years. He mentioned cases where the Swaar and Baldwin, when root-grafted, had not borne in fifteen years. Other trees, budded from good bearers, had borne good crops in seven years. He had never found a productive Rhode Island Greening tree, when root-grafted, at any age; but stock-grafted trees were always productive, when old enough for bearing.

Side or Summer-grafting. — This mode of grafting is known to very few persons; and yet it is superior to every other method. It is in a large degree free from the objections urged against cleft-grafting, and is so simple and easily executed, that the merest novice, armed with a knife, string, grafting-wax, and a piece of soap, can graft a tree as well as the professional propagator.

When a cion is to be set on the side of a branch, or small tree, make a slit on the stock, precisely as when a bud is

Fig. 4.

to be inserted, as represented by the illustration (Fig. 4). Next, commencing at the top of the slit, raise the bark on each side, to admit the lower end of the cion, which should be prepared before the slit is made. The cion should be short, not more than two inches long. One bud on it will suffice. A long cion is more liable to get displaced than a short one. Shave off the lower end true, and with a sloping cut, as shown by Fig. 4 herewith given, and crowd the prepared end down in the slit, in the same manner

Stock and cion for side-grafting.

that a bud is inserted. The end of the cion should fit neatly. If the stock be small, the end of the cion may be hollowed out a trifle with a small, sharp gouge. It will be necessary to touch the edges of the slit made for the reception of the graft with a piece of soap; otherwise root-lice, which are in the tops of the trees in summer, will enter the wound, and prevent the union of the stock and graft. Now bind the parts together, using strong cotton twine for large branches, or, for small ones, strong, coarse, woollen yarn, or such materials as are used in budding. The accompanying illustration (Fig. 5) represents a stock just grafted in this manner. After the ligature is tied, cover the wound with a thin plaster of wax. Good cions, neatly fitted, if inserted at the time the annual layer of new wood is being made, will unite with the limbs

Fig. 5.

Side-grafting.

as readily as buds. One of the eminent advantages of side-grafting is, that a new branch may be started on the side of a tree, or limb, where there are no branches, and where it is desirable to fill up with a limb any naked space, for the sake of symmetry.

Another Mode of Side-grafting is performed by cutting

Fig. 6.

Crown side-grafting.

off the branch, or the stock to be grafted, the same as for cleft-grafting. Then, instead of splitting the stock, make a cleft, or recess, for the cion (Fig. 6), with a back-saw, as represented below (Fig. 8), which has a thin blade attached near the ends of the saw, for shaving off the sides of the saw-kerf true and smooth, as fast as the saw enters the stock. Such a saw may be satisfactorily employed for grafting grapes and other kinds of fruit. It matters not how the cleft is made, if the inside bark of the cion fits neatly, with a gentle pressure, to the inside bark

Fig. 7.

A cion formed for crown side-grafting.

of the stock. When a stock is two or more inches in diameter, six or more cions may be set around the edge without splitting the end. The ends of the cions (Fig. 7) must be fitted true with a sharp knife, and be pressed into the cleft firmly. After the cions are set, apply a coat of grafting-wax that will not melt and run down in hot weather. Success in side-grafting will always depend more on skill in fitting the cion to the stock than in any thing else. Side-grafting also possesses this very important advantage over cleft-grafting: the cions, when set in a large stock, are frequently crushed by the

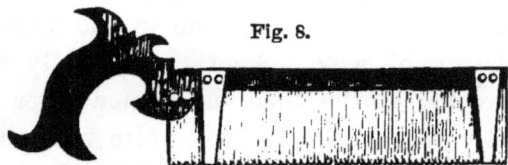

Fig. 8.

A grafting-saw, with blades for shaving out the sides of the recess like Fig. 6.

jaws of the stock, so that there is no vitality left in the part of the cion within the cleft. Large numbers of cions are destroyed by this means. When cions fail in cleft-grafting, rain enters the fissure, the cleft opens, and the end of the limb decays, to the serious injury of the tree. When fitting cions for this kind of side-grafting, it is important to pare the end of the stock smoothly with a sharp knife; then make a shoulder on both sides of the cion and cut and fit until the shoulder on both sides will set down tight to the inner bark of the stock. A cion will rarely fail to grow, if there is a good fit at the shoulder.

Dovetail Side-grafting.—I have never known a more successful mode of grafting both apple and pear stocks than this, which is more particularly adapted to large stocks with thick bark than to small ones. The stock is sawed off squarely, the end shaved smoothly, and dovetailed gains cut in the bark for the cions, as shown in the illustration (Fig. 9), representing the cleft or gain in the bark. A narrow strip of bark, say one inch long by one-sixteenth or so wide, is removed. Then, as the knife-blade is guided by a small straight-edge, to aid in making a straight cut, the edge of the bark is bevelled on both sides of the gain, so as to give the cleft a dovetail form for holding the cion. The width of the gain or cleft should depend on the size of the cion. Fit the lower end of the cion, as represented by the illustration, by cutting away about one-half true and smooth, with a square shoulder. Then shave off each side, true, down to the inside bark of the cion. If the gain be too narrow, dress it a trifle wider. In case a gain is made too wide, prepare

Fig. 9.

Dovetail side-grafting.

Fig. 10.

A stock crown
side-grafted.

another cion. The cion should fit so neatly that the end may be pressed down into the gain as tightly as practicable without lifting the bark of the stock. See that the end of the cion fits well to the bottom of the gain, and also that the shoulder sets down tight on the end of the stock. Several cions should be set in one large stock, like Fig. 10. If the grafting is performed when the cambium is abundant, almost every cion will live. Let the wounds be covered at once with a coat of wax. In most instances, it will be advisable to wrap several thicknesses of prepared ligament around the stock after a thin coat of wax is applied.

Spur-grafting.—This mode of grafting is practised but little. We have set cions according to this style; but they are not as likely to live as by side-grafting. Hence it is alluded to simply to show how grafts are set in this manner.

Procure a " firmer" chisel, say three-eighths of an inch wide, grind the edge in the form of a round-pointed spear. Drive this instrument perpendiculary down through a growing root of an apple-tree, withdraw it, crowd in a cion having a wedge-shaped end, cover the wound with wax, and the cion is ready to grow. The tool must be sharp and smooth, so as to make a clean cut; and the cion must be crowded in quickly, before the fissure closes. Let the tool be driven into the limb of a tree in the same manner, and a cion inserted. Cions should be set in this manner when the cambium is abundant. But even under the most favorable circumstances, side-grafting will be found preferable to this method.

Cleft-grafting.—The usual manner of grafting stocks from one to three inches in diameter is illustrated by Fig. 11, p. 33. The stock, *b*, is first sawed off with a fine-toothed saw,

like Fig. 8, when the end of the stock
is shaved smoothly, so that one can
readily see the point of union between
the bark and the wood. The blade of
the grafting-knife, shown below (Fig.
12), is then driven into the stock, *b*,
to split it, as represented. After the
stock is split, the chisel, or wedge of
the knife, is driven in the middle of the
cleft, to hold it open while the cions, *a*,
are being set. The blade of Fig. 12
should be about six inches long, besides the handle. The heavy part,
which constitutes the head of the chis-

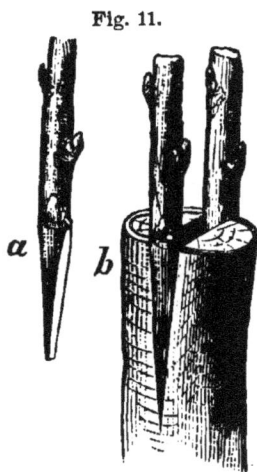

Fig. 11.

a *b*

Cleft-grafting.

el, should be about three eighths of an inch square, and the
chisel should extend about one inch below the shoulder.
The chisel at the base should be one-eighth of an inch thick,
which will be sufficiently thick to open any cleft, except for
a very large stock. This combined tool will be found more
convenient than to have the chisel at the end of the knife,
as it will maintain the balance with less difficulty. The

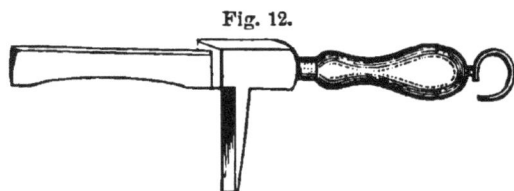

Fig. 12.

Grafting-knife and chisel.

blade should be one-
eighth of an inch
thick on the back.
A wire hook is pro-
vided in the handle
for suspending the
tool from a branch

when a person is grafting the top of a large tree. The cions, *a*, are dressed off true, like a long wedge, and inserted
so that the point of union between the bark and wood, on
both the stock and cions, will exactly coincide. A sharp
knife should be employed for making cions, and the ends
should not be dressed too blunt nor too slim to fit the cleft.

2*

Fig. 13.

Cleft held open
by a wedge.

Many times the stock will close so firmly on the cions, as has already been stated, as to crush the ends. The grafter, when removing the chisel, must watch the pressure of the sides of the stock on the cions; and if the ends are liable to be crushed, a small wedge, as shown in Fig. 13, must be driven in by the side of the chisel, so that the cions may not be injured. Grafting-wax is now applied, as represented by Fig. 14, by pressing it into all the crevices, and covering the wound between the grafts. Great care must be observed, after cions are set, not to move them. Fig. 15 represents a small stock cleft-grafted, with only one cion, made with a shoulder, as shown by Fig. 16.

Fig. 14.

Grafts and the
stock covered
with wax.

The cion should be set, and taken out, and fitted with a sharp knife until each shoulder will set down to the stock watertight. Then the cion will scarcely fail to live.

Figure 17 represents the manner in which small trees are often grafted. A cion, *b*, about two inches long, is fitted, as shown, to the stock, *a*. If the cuts are made true and smooth, and the two parts be properly united, a graft seldom fails to grow.

Fig. 15.

Fig. 16.

Cleft-grafting, with a
shouldered cion.

Fig. 17.

Splice-grafting.

The cion is bound to the stock with grafting-wax, which will hold it in the proper position until a union is effected. The wax for this purpose should be tempered with a larger proportion of resin, so that it will be hard after it is applied to the wound.

By the different parts of Fig. 18, American whip-tongue-grafting is represented in the various stages. The parts

Fig. 18.

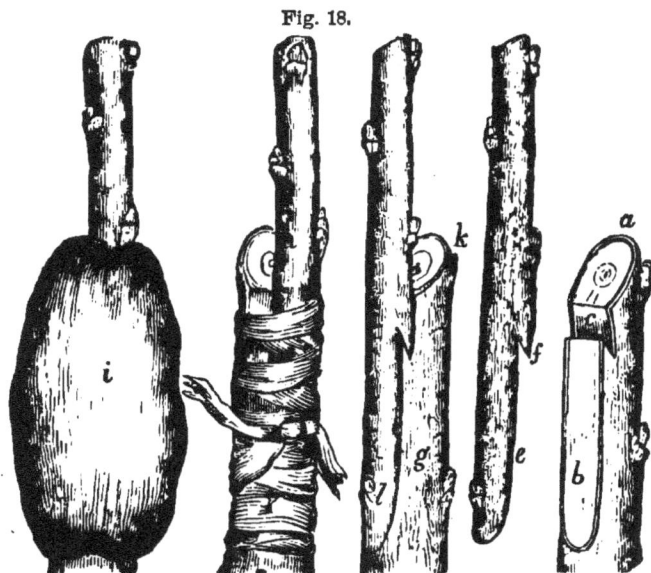

American whip-tongue-grafting.

shown at *a*, *b*, *c*, represent a stock prepared for the cion, *f*, *e*. At *k* and *l*, the stock and cion are united. At *h*, the ligature is applied. At *i*, the work is shown covered with grafting-wax. After the graft has begun to grow, a sharp knife should be drawn through the back part of *i*, to cut the bandage. Then the growth of the wood will not be obstructed. It is eminently important that every stock grafted in the foregoing manner should be examined in ten days after the cions are set, for the purpose of re-grafting, where the cions fail to grow, and also to release the ligature. When stocks are very thrifty, the grafts will frequently expand, in a few days, to such an extent as to bury

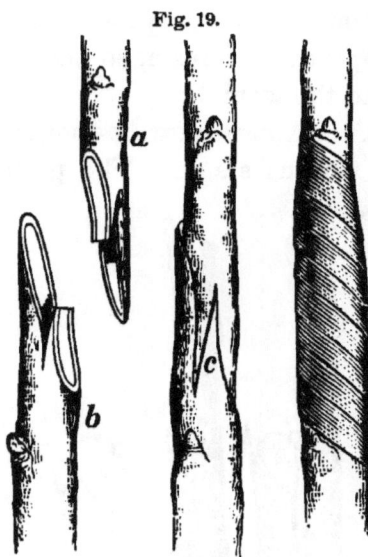

Fig. 19.

American whip-tongue-grafting.

the strands of the twine into the tender bark, to the serious injury of both stock and graft.

Fig. 19 represents another plan of American whip-tongue-grafting, in which a is a cion, b a stock. At c the cion and stock are united. At d the union is wrapped with a narrow strip of cloth saturated with grafting-wax. Figure 20 represents still

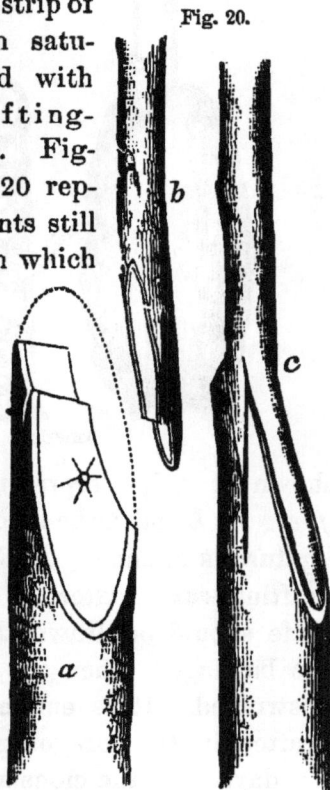

Fig. 20.

Whip-grafting large stocks.

another style of whip-grafting, in which a is the stock, b the cion, and c the two united, ready for the wax.

Before a person attempts to graft valuable stocks by whip-grafting, he should procure a bundle of twigs and a sharp knife, and spend one or two hours during his leisure evenings in educating his hands to handle a sharp knife with skill and precision, in making "neat fits" between the stock and cion. It will be found a great convenience, also, to employ a small stick having a groove on one side, into which the stock, or the cion, may be placed when the end is being dressed off.

At Fig. 21, on the following page, a style of saddle-grafting is shown, which will require no explanation, as the illustrations show how the fitting of cion and stock is done. Fig. 22, below, represents the manner of saddle-grafting stocks that are larger than the cions. The butt-end of each cion is split, shaved, and made to fit neatly to the stock, as shown. In some places, one piece is held down to the stock by means of small nails, or upholsterers' trimming-tacks. This latter style of grafting is practised but little; and it is represented here more for the novelty than for its utility. One great advantage

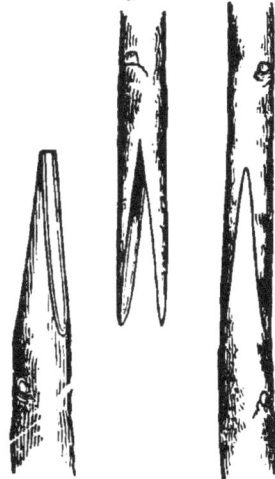

Fig. 21.

Saddle-grafting small stocks.

of setting grafts like Fig. 21, is, no bandage will be required to hold the cion on the stock. Good wax is all that will be necessary. When stocks are grafted where they sprang from the seed, and where the apple-trees are to grow, if the cions are fitted neatly, like Fig. 21, they seldom fail to grow rapidly.

At Fig. 23, represented on the following page, the tops of two small trees are brought together, and the side of each one shaved off true, so as

Fig. 22.

Saddle-grafting large stocks.

Fig. 23.

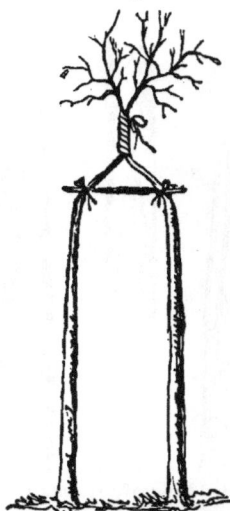

Grafting by approach.

Fig. 24.

Inarching, or grafting by approach.

to make a close union, when the parts are bound securely and covered with wax, and the tops are held apart, and in the desired place, by a stay bound from tree to tree. At Fig. 24 the manner of inarching is represented, which will require no further elucidation than the cut will give. Inarching is frequently employed to connect the ends of two branches above the fork of a valuable tree, to prevent one or both sides from being split down by a furious wind. The two parts to be united must be held by means of stiff splints of wood, until the union is perfect and strong.

Inoculation, or Budding.

In that smooth space a narrow slit we make,
Then living buds from bearing trees we take:
Inserted thus, the wounded bark we close,
In whose moist womb the tender infant grows.

DRYDEN's *Virgil*.

The operation of budding consists in cutting a bud from a young twig and inserting it beneath the bark of another branch, or in bringing the bud and stock together in such a manner that a union will readily take place between the two. The future tree and fruit will always partake of the character of the tree from which the bud was taken.

Satisfactory success in budding will always depend on several contingencies; such as, 1. The condition of the stock and bud. Budding can not be performed successfully when the bark of the stock will not peel easily. Buds must be set when there is a bountiful supply of *cambium*, or *mucilaginous material* in a semi-fluid state between the bark and the wood of the stock, to facilitate the vital adhesion of the bud. By making a slit in the bark of the stock, one can readily perceive whether the cambium has become too hard to unite with the bud, or whether the supply is still insufficient to effect a union. 2. Varieties of the same species are always more successful than a bud taken from a branch which is not closely allied to the stock. Apple-buds should be inserted on apple-stocks, pear-buds on pear-stocks, and peach-buds on young peach-trees. 3. The bud should always be taken from a shoot of the present or the preceding year's growth, and inserted in a stock not over two years old. Buds are seldom inserted in stocks over one year old. If buds are set in the latter part of summer, a stock of the present season's growth is chosen. 4. It is important to take such buds as have come to full maturity, as a half-mature bud will often fail to produce a shoot, even were it to live. We have sometimes inserted immature buds, which have adhered to the stock, but which pushed out no stem for two years after. 5. The inner surface of the bud must be fitted so neatly to the corresponding surface of the stock, that a union can not fail, provided the operation is performed at a proper period in the growing season. 6. Mechanical skill and dispatch are essential to satisfactory success. The incision should be made neatly; the bark lifted a trifle, without disturbing the cambium beneath it; and the bud should be cut from the twig, inserted, and the ligature applied in a few seconds.

The best Time to Inoculate.—The best period of the year to bud must always be determined by the growing stocks.

When the cambium is in a semi-fluid state—whether in spring, summer, or autumn—insert the buds. In this latitude, most buds are inserted during some one of the summer months, or in September. In stocks that continue to grow late in the season, buds are frequently inserted in September, and some in October, but August is preferable.

Cause of Failure.—Beginners are sometimes surprised to find, after budding a lot of thrifty young stocks, that almost every bud has failed; and they are quite at a loss to account for the failure. But experience shows that the failure is caused by the over-luxuriance of the stock, and the thin, watery condition of the sap. If the operation had been deferred until the sap had thickened, the result would have been reversed; and instead of only one in a hundred succeeding, there would have been only one per cent. of failures. The cherry is more liable than any other sort of fruit to " drown out " the bud, as it is called. Hence the best time to bud cherry-stocks is just as soon as they begin to slacken their growth and show a yellow leaf here and there. If this time is chosen, and the work done skillfully, there need be little fear of failure. Trees which are not growing vigorously should be budded early. As soon as wood sufficiently ripe to furnish buds can be found, if the bark on the stocks will peel, it will not be too early. When a large quantity are to be budded, the work should be taken in hand early, so as to get through in season, commencing with the least thrifty.

The object of budding is the same as of grafting, viz., to propagate a desirable sort of tree or plant. The only difference between a bud and a cion is that the latter is a development of the former. Fruit can generally be obtained by grafting from one to two years sooner than by budding. But when a variety is very rare, we can, by budding, get new limbs from single eyes; whereas, in grafting, we have to use three or four eyes. Some trees, moreover, propagate more readily by budding than by grafting. The stone-fruits exude so much gum, when grafted, that it is hard to succeed in the work. Then, too, in all kinds of fruit where grafting has failed, or been forgotten in spring, budding may be resorted to in summer. Then, if budding fail in summer, the same stocks may be grafted the succeeding spring.

As has been stated, the shoot from which the buds are taken must be of the current year's growth, and must be mature. This maturity will be shown by the forming of buds at the axils of the leaves, and of the terminal buds. The best buds for working will be found along the middle of the shoot. If it is necessary, in order to have the buds ready to meet the growth of the stock, that the scion or branch from which buds are to be taken should be made to hasten its maturing of the buds, pinch off the ends of the shoot one or two weeks before the buds are to be set. In from eight to twelve days the remaining buds will have ripened and fitted themselves for forming new branches. If this pinching is done early in the season, and the branch left on the tree, the result is, the buds, after ripening, send out new branches, and make a sort of second growth. The accompanying illustration (Figure 25) represents the correct way of holding cutting off a bud. when cutting the

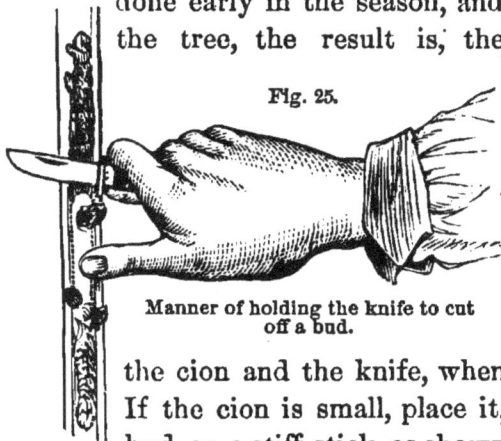

Fig. 25.

Manner of holding the knife to cut off a bud.

the cion and the knife, when If the cion is small, place it, bud, on a stiff stick, as shown in the diagram. Then a bud can be cut off with far greater precision than if no stick were employed. Never attempt to cut off a bud with a dull knife. Let the blade be ground to an edge; then whet it on a fine-gritted oil-stone, so that the cut can be made true and smooth. If a bud be haggled off with a dull knife, and the surface is not straight and even, the union will be quite uncertain. When cutting off a bud, employ a *drawing* cut of the knife.

Fig. 26.

A cion for budding.

Cions for Budding.—Always select the terminal shoots for budding. Then, as soon as the shoot is cut from the tree, let every leaf be cut off, as represented by the figure, as the leaves rapidly exhaust the liquid in the bud when its connection is separated from the root, and thus impair vitality. The buds of the upright shoots of a tree are said to make more vigorous growth than buds from lateral shoots; and buds from bearing trees are said to form fruit sooner than buds from young trees.

A good Budding-knife is all-important, as one can not use a large pruning-knife for removing the buds advantageously. A knife with a thin blade, rounded at the point, will be found most convenient. The most important consideration is, to

Fig. 27.   **Fig. 28.**

A pair of budding-knives.

have a thin and narrow blade with a keen edge. A rough-edged razor is no more certain to make a painful shave than a rough-edged budding-knife is to make an unsuccessful bud. It requires a good knife, a steady hand, and considerable practice to cut off buds handsomely, well, and *quick.* As to taking out the particle of wood attached to the bud, it matters little, if the cut be good and not too deep. In taking out the wood, great care is necessary to avoid taking the root of the bud with it. Then, when the bud is in its place, it must be well tied up. Nice, smooth, soft strips of corn-husks applied wet, like narrow ribbons, are the best and most convenient in common use. Every part of the cut must be wrapped so firm as to exclude air completely; and this should be done as quickly as possible, as the air soon blackens the inner surface of the bark, and prevents the perfect union of the new parts that are placed in contact.

Different Styles of Budding.—There are in vogue three styles of budding, all of which are substantially the same. The difference will be perceived by the following illustrations (Fig. 29, p. 44), each of which will give the beginner so much of an insight into the process of inoculation, that even young boys and girls may bud rose-bushes or young fruit-trees with satisfactory success. The ordinary process of budding, denominated *Shield-budding* or *T-budding,* is represented by the accompanying designs, of which *a* is the representation of the stock, and *b* is the bud. The following will furnish a correct idea of the manual operation: After selecting a smooth place on the stock, *a,* make a cut with a sharp, round-pointed knife, in the form of the letter T, about one inch long. Be careful, when cutting, to press the edge of the blade only through the bark into the semi-fluid cambium, and not into the wood of the stock. Now lift the corners of the bark with the knife; then cut a bud,

Fig. 29.

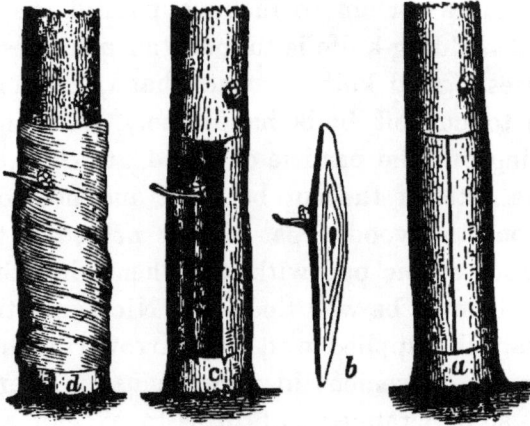

Shield-budding.

b, from the twig (Figure 25), as shown by the line around one bud, and thrust it carefully down into the sheath, *c*. If the piece to which the bud is attached be too long, cut off the upper end, so that the ends will fit closely to the bark of the stock, *c*. Now pass a ligament around the stock, both above and below the bud, as at *d*, and tie the end securely. The small piece of wood that is cut off the twig may be removed if it will separate easily. But it will be quite as well to allow it to remain.

The terminal bud of a twig is sometimes inserted, instead of a side-bud. The terminal bud may be employed, if a neat fit is made between stock and bud, more successfully than if a side-bud were used.

Fig. 30.

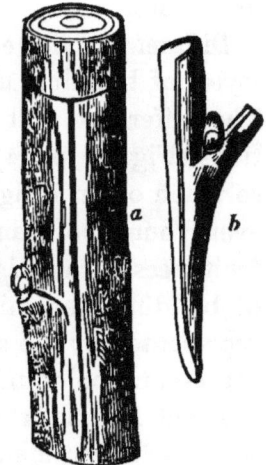

Stock, *a*, and bud, *b*, shown separately.

Fig. 31.

Annular budding.

Fig. 31 represents the manner of performing *annular budding*, or *ring-budding*, which is done by taking a piece of bark, say three-fourths of an inch long, on which there is a good bud, from a twig a little smaller than the stock. Then remove a piece of the same length from the stock, and wrap the piece that bears a bud around the stock, and secure it with a soft and elastic ligament. Young stocks of nut-bearing trees, that have thick bark, are frequently budded in this way with more satisfactory success than by the T-style of budding. Cotton and linen ligaments are objectionable for closing the lips of the bark, as there is but little elasticity in such materials. The ligaments should be somewhat elastic, so that the strands may yield a trifle, as the stock enlarges by growth.

Management of Buds.—After the buds have been set about six days, each one should be examined by some competent and careful person who will exercise proper judgment in removing the ligaments from the buds that have adhered firmly, and in loosening the ligaments around others, when the strands are so tight as to form creases in the bark. When the lips of the T-cut have not united with the cambium, so as to hold the bud, the ligament should remain on longer. As soon as the bud has united firmly with the stock, the ligament should be removed. Buds that are set in the latter part of summer are not expected to send up a stem until the next spring. Then, early in the growing season, the stock should be cut off about one-eighth of an inch above the bud, not square across, but a little slanting, so that the wound will heal soon. Grafting-wax should be applied to the wound as soon as the cut is made, that the wood may not dry up, to the injury of the bud. If shoots from buds push upward too rapidly for their strength, a small stake should be set near the stock, to which the tender stem should be tied with soft shreds of old cloth, to prevent the wind from breaking the young shoots off; or the shoot may be secured like Fig. 32, represented on page 46. In many instances, the stock is cut off so near the bud that, if

Fig. 32.

Manner of tying
the shoot, *a*, to
the stock, *b*.

grafting-wax is not applied, the wood of the stock dries up so rapidly that the young shoot withers and dies.

Budding vs. Grafting.—Budding has some advantages over grafting.—1. Budding requires less skill than grafting. Consequently, a young beginner may insert three or four buds in one small stock, with the assurance that one or more will live. Whereas only one cion can be set in a small stock. 2. In case four buds were to fail, the operation may be repeated every ten days without checking the growth of the stock, which is impracticable when cions are set in by grafting. 3. Varieties of fruit can be budded with satisfactory success which can not be propagated by grafting. In this latitude pomologists rarely attempt to graft the peach. But we have seen it stated that in the latitude of Georgia there is no difficulty in grafting the peach successfully. Many other trees and bushes may be budded which can not be grafted with success. But *old stocks* can be grafted in which buds would never live if they were inserted. In such instances, the process of grafting is superior to budding. A young tree may be sawed off at the collar of the stem and cions inserted, which will often grow two to four feet high before winter. In case the graft should fail, sprouts would probably start, which could be budded in August. Hence the eminent advantage of understanding both operations, and knowing how to employ either process advantageously when the other fails. If branches of a large tree were grafted, and the cions should not live, the young sprouts, which will usually start at the end of the stub, may be budded in August. Many beginners who did not understand how to take advantage of a failure have, in consequence, lost sev-

eral years' growth and value of their trees. Our own practice always has been to bud young stocks in August or September. Then, if the buds failed, the stocks were grafted
the next spring close to the ground.

Management of Apple-trees in Nurseries.—The following
constitutes the practice of many nurserymen: They collect
apple-roots about as large as goose-quills, keep them in
sand in a cellar, cut them in pieces about four inches long,
graft each piece, during the winter, within doors, pack the
grafted stocks in bundles, and plant them out in rows in
the spring where a nursery is to be formed. Some plant
them thus: Thrust a spade into the soil, so as to have the
spade extend equally on both sides of the mark for the
row; now push the spade from you, and then withdraw it.
This operation will leave a wedge-shaped hole ten inches
deep, seven long, and about two broad. It is necessary to
have another man to put in the grafts; he should put in
two, each two inches outside of the mark left as a guide,
holding them till the first or spade-man repeats the operation with the spade, which will be three to four inches
from the edge of the first hole. In making the second, the
soil is pressed against the grafts in the first; and thus the
operation not only makes the new hole, but closes the last;
and while the spade is being withdrawn, the second man
gets the grafts for the new hole. There is no time lost
from one operator waiting for the other. Set only half an
inch above ground. The young trees are then cultivated
in the same manner as a crop of carrots. All weeds are
kept down, and the ground is kept mellow and loose. After they have grown an inch or two they are sprouted,
which consists in taking all the sprouts off but one, the
best. In the fall they should all be taken up and stored in
the root-cellar; and during the winter, trim the roots and
top, and cut the top to within an inch of where it started;

the roots should be trimmed small enough to enable you to
plant them in the nursery-row with the spade. The second year plant these yearlings in the nursery, the same as
the year before. Rows four feet, plants one foot apart in
the row, leaving an inch above ground. Sprout as before.
In the fall and during the winter, trim them to *whips*, leaving only one straight shoot. The third year let them grow
as they may. In the fall and winter, trim them to *whips*
and top them at four feet. The fourth year, during summer, trim off any limbs that are too low on the trunk to belong to the head. In the autumn the trees will be ready to
transplant. Some nurserymen transplant their young trees
but once. Such a practice will produce young trees of fine
appearance; but we do not recommend this mode of producing an orchard. If we were to plant a hundred orchards,
the trees should never be produced in such a manner. Let
the seed be planted where the trees are to grow. We are
well aware that nurserymen will hoot at this, because it en-
dangers their business of selling trees. But such trees can
rarely be relied upon, any more than a herdsman can depend on the veriest scrubs of neat cattle for superior animals, simply because they have not been bred according to
the requirements of vegetable physiology. This is the
most economical way to produce apple-trees *to sell* to people who do not understand the difference between a valuable, hardy, and prolific tree, and one that is worthless. Not
one-fourth part of the seeds from which the thousands of
fruit-trees in nurseries have sprung were any more fit for
producing valuable trees than the half-ripe and shrunken
kernels of wheat and other grain are suitable for choice
seed.

The True Way to produce Fruit-trees.—A beginner may
listen to the talk of those who have fruit-trees to sell; and
yet, if he desires to obtain trees that will supply him with

fruit, and be a choice heritage to his successors, let him begin *right* by selecting the seed from apples or pears with his own hands; plant them where the trees are to grow; bud the young trees; and train and cultivate them for a few years, until they have obtained sufficient size to require but little more care.

Now, then, what does a beginner desire to accomplish? What *end* has he in view, near or remote, in the future? Why, simply, the object to be attained is, *hardy*, *thrifty*, *productive* trees, which will not fail to yield fair crops of excellent fruit every season. There is but little difficulty in accomplishing all that may be desired, if one can find a hardy fruit-bearing tree in his vicinity. It is assumed that the *stock* of a fruit-tree will exert a marked influence on the production of the fruit with which it may be grafted. The Rhode Island Greening is a fair bearer in all sections of the country, where the tree has not been *starved*. The English Streaks and the Romanites are also hardy, and *naturally* prolific. Select a few of the fairest apples of these varieties, or the seed of any other hardy variety, and plant only the largest and most perfect seeds from the fruit. In some of the specimens there may not be a single seed fit to plant. In others, one seed only can be found. The same principle will hold good with pears or any other fruit. None but the *best* seeds must be selected. The fruit of any pear-tree that is hardy, and has produced a crop every season for several successive years, may be selected, from which to obtain a supply of seed for raising young pear-trees. As soon as they are removed from the fruit, before the kernels have been allowed to dry, mingle them with sand a little moist, and keep them in a cool cellar until cold weather. Then plunge the box in the ground, so that the seeds will freeze. Early in the growing season, stake out the ground, which is supposed to be as mellow

as a carrot-bed, run a crowbar down four feet into the
earth where each tree is to stand, make a large hole, fill it
with rich soil, and plant two or three seeds about one inch
deep.   Stick the seeds point downward, so that they will
come up without difficulty.   Cover them with fine loam.
The seeds should not be planted more than one inch apart.
If they all grow, the best stem only should be allowed to
stand.   A strong stake should be driven into the ground
before the seeds are planted; and the seeds should be stuck
in about six inches from the stake, on the south side.   The
object of the stakes is to protect the young trees.   If the
soil is sufficiently fertile to yield fair crops of grain or potatoes, the tap-root of every young tree will strike four feet
into the earth the first season; and the tops "will grow
like sparagrass and spread like applesas."

As soon as the young trees are large enough, they should
be inoculated with buds taken from the topmost branches
of trees that always bear a bountiful crop.   If the land be
kept clean, and if the surface or coronal roots are not mutilated and torn from the stump, every tree, at the end of
ten years, will have attained a height of over twenty feet,
and will be loaded with fruit; while many ordinary nursery-
trees, planted in the usual manner, will never yield a fair
crop.   If an orchard is produced in this manner and receives proper care, the trees will yield bountiful crops for a
hundred years.   (See *illustrations*, p. 22.)

Why every Farmer should produce his own Trees.—When
a person *purchases* an apple-tree, he has no assurance that,
if it ever produces fruit, the product will be the variety
that he bargained for.   Neither has he any assurance that,
if the tree bears, it will yield even one-fourth part of a crop.
Stocks of young trees are often produced from the poorest
seeds of a very poor growing tree, a shy bearer, and a worthless variety.   Then, if the cions of a shy bearer and a ten-

der variety be worked on inferior stocks—as they often are
in large nurseries—what can be expected from the tree?
There are untold numbers of apple trees of this character
all over our country; and they can never be made to yield
abundant crops, even if the tops are regrafted and the soil
renovated. It is well known that many tree-peddlers are
not over-scrupulous in their business. They will sell a person any variety of fruit-trees he may desire to purchase,
whether they -have such trees in their possession or not.
This is frequently done. Tree-peddlers have told us that
they have often sold any variety of apple-trees that they
had in their -possession, for *other* varieties that were called
for. They received their price for the spurious trees, which
was all they cared for.

It will usually require six to ten years to determine an
error, or trick, or fraud, in the purchase of fruit-trees. And
even then, no person who hates strife and the uncertainties
of legal contests would undertake to ferret out a swindler
in the purchase of fruit-trees; but there may be the trees,
after ten or twenty years, comparatively worthless—cum-
berers of the land. There are but few orchards in the
country in which more or less worthless trees can not be
found.

Again, even when excellent apple-trees are ordered of a
reliable nurseryman, it often happens that all the small
roots are thoroughly dried up and killed before the trees
are transplanted where they are to grow. Tree-diggers
are frequently ordered to take up several thousand trees
with a horse-digger, leaving hundreds of them lying in a
hot sun, and exposed to drying winds for half a day, or
even longer. Then, before the roots are wrapped in moss,
there is not a vestige of vitality in many of them. In many
instances, a car-load of fruit-trees is shipped several hundred miles, when they are tumbled into a hay-rigging and

carted about town all day, exposed to drying winds and
sunshine, and frequently to cold and frosty nights, which
will destroy the vitality of every root. We have often
travelled on steamboats and cars, where we have seen fruit-
trees without any protection, piled in the open air, where
the roots have dried to death in a few hours.

A farmer purchases, for example, a supply of apple-trees
to be sent one or two hundred miles. As orders at the
nursery may be large, the operation of digging must commence early in the season. The trees are dug up before
the frost is really out of the ground; and before they can
reach their destination, they are often frozen and dried, alternately, for two weeks. The great wonder is, that those
who attempt to produce orchards with purchased trees succeed half as well as they do. But immense numbers of
failures in all parts of the country show conclusively that
there are grave faults somewhere.

Another objection to nursery trees is the fact that, in
many nurseries, the young trees have been forced into an
unusually large and tender growth by frequent applications
of stimulating manures. The nurseryman produces trees
to sell. He has no further concern than to prepare for the
market such trees as will supply an active demand at an
exorbitant price. Hence he piles on the manure, and produces in the shortest possible period the largest possible
growth. He will not be responsible for the results after
the trees have been transplanted into an orchard. Beautiful young trees removed from a nursery, where the soil is
as rich as a fertile carrot-bed, to land of ordinary fertility, or
to a poor soil, will usually receive a "set-back," or "stunt,"
from which they seldom recover.

The foregoing suggestions will furnish sufficient reasons
for starting young fruit-trees of any variety of fruit on
such ground as may be chosen for the orchard, that they

may not be checked in their future development. Hence one will always be more certain of having hardy and thrifty trees when they are obtained from a nursery where the land is in rather a poor state of fertility than from a nursery where the soil is rich. But, after all, the correct way is to plant the seeds where the trees are to grow. By adopting the plan herewith recommended, a person can produce a fruitful orchard much sooner than by purchasing his trees.

Practical Operations.—In 1843 we received the catalogue of a nurseryman who was recommended to be "thoroughly reliable as to the genuineness of every tree that was ordered from his nursery." His apple-trees and pear-trees were represented in his catalogue as being very large and fine— "four to five feet high"—hardy and thrifty, and would be shipped for fifty to seventy-five cents each. As we desired to start an orchard, we forwarded the money, and gave an order for the trees early in the spring. After the season for planting trees had so far advanced that we had thought it quite too late for transplanting, our trees arrived. But, instead of being thrifty and large, suitable for transplanting, some of them, for which we sent seventy-five cents each, were only one year old, and some had only been budded the previous season. They had been exposed to the air for so long a time, that it was only by the best care that life was preserved, without one inch of growth, till the next season. Many of our neighbors were treated in the same manner by the same nurseryman. And yet we knew him for more than twenty years, up to the day of his death, as "a reliable nurseryman!"

Prepared Bandages for Budding and Grafting.—Cut cotton cloth, such as sheeting, into narrow strips, say half an inch wide, and sew the ends smoothly together in the same manner as carpet-rags are prepared. Then wind the long strip on a small stick loosely, and in a diagonal direction, so that a portion of the roll will not fall off the stick without unrolling. Prepare one or more bunches of an oblong form, as large as a man's fist. These shreds may be made also of under-garments that are worn out; or a yard or two of light shirting may be cut into strips one-fourth of an inch wide and sewed together. Now put one pound of beeswax, one pound of resin, and one and a quarter pounds of tallow into a deep vessel, and melt it by a gentle heat. Then plunge the balls of bandages into the boiling liquid, holding them below the surface until the liquid has forced all the air from the interior of the balls and saturated the cloth. This may be determined by observing when bubbles of air cease to rise from the surface of the liquid. By this means grafting-wax is simply applied to a bandage in a very economical manner; and the bandages are in a convenient form for use at any time and for all kinds of grafting and budding, as it will not unwind of its own accord. In cool weather, if the wax with which the roll is saturated should be too hard, keep it in a vessel of warm water while using it, so that the wax will be sufficiently plastic to work. Such bandages require no tying, as the ends will adhere to the stock or bandage. Furthermore, there will be sufficient elasticity in such bandages to allow the stock to expand by its growth, and no cracks will be formed in the wax, as there is when nothing but the clear wax is employed. After the bandage is applied to the stocks, a little wax may be spread over the end where the saturated bandage does not cover the wound. The width of such bandage-material should be increased for large stocks.

Another Way of making Grafting-wax. — Any smaller quantity may be made by observing the correct proportions. Add tallow to make it softer, and resin to render the wax harder. Take six pounds of resin, two pounds of tallow, and two pounds of beeswax; pulverize the resin first, and put into a clean, dry iron pot over a slow fire; stir constantly, until it is all dissolved; then add the tallow and wax, and stir the mixture until it is melted; pour the mass into a vessel containing cold, clean water; commence immediately at the edge of the wax, and pull pieces of it, as you would pull molasses-candy, rubbing your hands first with tallow; and continue to do so, now and then, until it is finished; but work it no longer than is necessary to take the water out. Separate it in rolls six inches long, and a little thicker than a candle. Put these rolls on a dish or pie-pan in the cellar until wanted for use. They will keep for years. One day before using, hang them in the kitchen, not near the fire, and they will become · pliable. Always have a piece of tallow near, to rub the hands while using the wax, as tallow will prevent it sticking to the fingers. Less beeswax and more tallow may be employed, if desirable. Grafting-wax may be made without beeswax; still, a small proportion of beeswax will render the mixture much more valuable.

Liquid Grafting-wax.—Figure 33 represents a vessel for

Fig. 33.

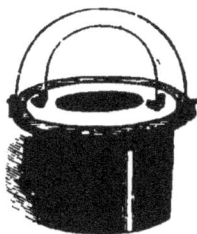

Pot for grafting-wax.

containing liquid grafting-wax, to be applied with a small paint-brush. The large outside vessel may be of cast-iron or tin. A common tea-kettle will subserve the same purpose. A small tin pail, to receive the wax, is placed in the opening of the kettle. At the top of the wax-kettle there is a broad flange to support it. The wax in the small kettle is heated by the hot water in the large vessel, without danger of burning it.

CHAPTER II.

PREPARATION OF THE SOIL FOR AN ORCHARD.

Now, long before the planting dig the ground
With furrows deep, that cast a rising mound;
And hoary frosts, after the painful toil
Of delving hinds, will rot the mellow soil.—DRYDEN'S *Virgil.*

THE manner of preparing the ground for an apple-orchard will depend very much on the character and condition of both the soil and the subsoil. The whole ground for an apple-orchard should be pulverized twenty inches deep, so thoroughly that the roots of the young trees will spread rapidly through the entire seed-bed. A person who is about to plant an orchard must exercise his own judgment in regard to deepening the soil, as the ground in many places is so porous and mellow that roots of growing apple-trees will strike down six or eight feet. Where a person can thrust a spade or shovel down, without difficulty, through the subsoil, all the preparation requisite will be simply to plough, manure, and scarify the surface-soil, as the ground is usually prepared for a crop of carrots or onions. But where the compact subsoil extends up to the second rail of the fence— as it is said to, along the slopes of some of our Northern lakes and rivers—a great deal of work must be done before the ground will be in a suitable condition to receive the trees or the seed. It will pay well to perform this job in the most thorough manner, as it is a piece of work that is done for life. If not performed thoroughly, before the seed is planted or the trees are set, it must go undone. By a proper pulverization of the subsoil twenty inches or two feet deep, where the subsoil is compact, trees will grow

Fig. 34.

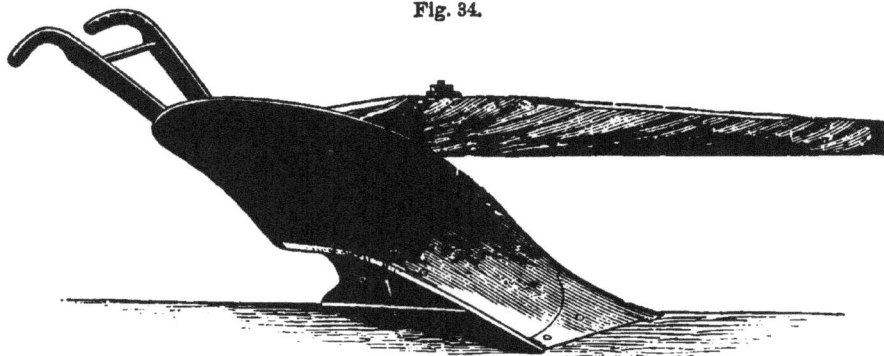

DR. GRANT'S "IONA GREAT TRENCH-PLOUGH."

The cut of a large plough, herewith given, Fig. 34, represents one of Dr. C. W. Grant's huge implements, which he ordered to be made by the "Peekskill Plough Company," of Peekskill, N. Y., for the purpose of pulverizing the soil to a great depth on Iona Island, where his celebrated vineyard is located. With a strong team of four to eight good mules, horses, or heavy oxen, such a plough will cut furrows, by running twice in a place, thirty inches deep. The implement is made very strong, and is well adapted to the purpose for which it was intended. Of course, where the substratum is full of boulders and bars of hard-pan, it would be difficult to draw any plough as deep as thirty inches; yet on many kinds of soil every square yard may be thoroughly broken to that depth.

larger in ten years than they would have grown in twenty years, if the subsoil had not been broken up so finely that the roots could spread deep and wide. Small, narrow, and deep post-holes in a compact subsoil will not answer. We want to put in the subsoiler beam-deep, this way, crossways, cornerways, and diagonally, so that every particle of the hard stratum may be broken up. It is always best, if practicable, to keep the thin stratum of soil or surface-mould on the top of the subsoil, especially where the under-stratum is compact, heavy, and less fertile than the surface-soil. The writer once prepared heavy land for an orchard by throwing the ground in high ridges with a three-horse plough. As there was no sod on the surface, a ridge was formed midway between the places for the rows, and the ground was ploughed several times, until a broad and deep middle furrow was produced where the trees were to grow.

3*

Fig. 35.

Figure 35 represents a strong two-horse subsoil-plough, which is designed for pulverizing the substratum beneath the surface-soil. Such implements are driven in the furrows of a common plough. The share and flange pulverize the hard ground, leaving it in the bottom of the furrows. A strong single team will draw such a plough.

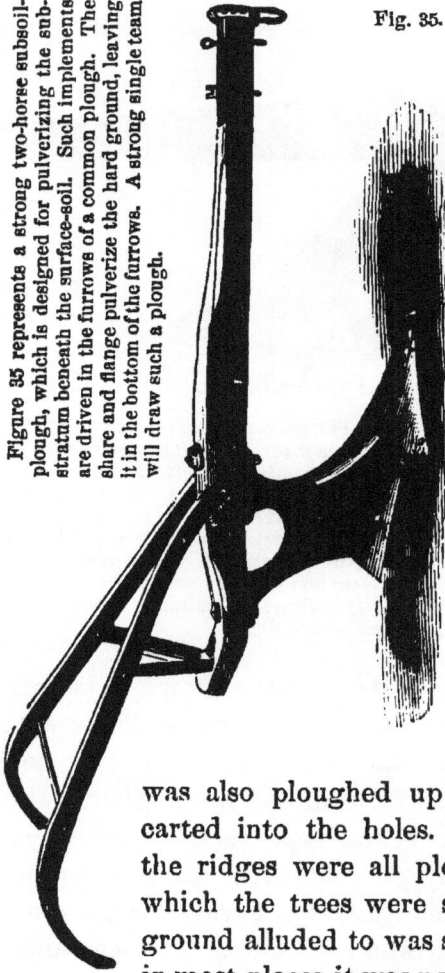

A TWO-HORSE SUBSOIL PLOUGH.

The ground was then staked out, when holes six feet in diameter were dug twenty inches deeper in the hard earth than the plough had been drawn. Mellow soil was then carted from another field, and shovelled into the holes. About half a wagon-load was deposited in each hole. As one man shovelled from the wagon, another returned the hard earth that had been removed from the holes. Rich turf was also ploughed up along the highway and carted into the holes. After they were filled, the ridges were all ploughed down level, after which the trees were set in their places. The ground alluded to was so compact and hard that in most places it was necessary to run the plough three times in one place, with a man on the beam, before we could work the implement down a foot below the surface-soil. Then, when we came to deepen the places where the trees were to be set, every inch of the earth had to be dug up with the sharp point of a digger's pick. Two men, by laboring faithfully, after the ground had been thoroughly subsoiled, were able to prepare the places and set out only ten to twelve trees in a day. But the rapid growth of the trees and fine crops of fruit assured us that such labor had not been misdirected. The truth is, roots of fruit-trees can not spread downward in such compact ground, except at a *very* slow and unsatisfactory growth, from year to year.

How to Plough Deep.

The plough with ill holding turns quickly aside.—TOM TUSSER.

When preparing the ground for an apple-orchard, where the substratum consisted of a compact calcareous, gravelly clay, the ploughing was done with a heavy yoke of oxen, and a span of horses forward of them, attached to a deep-tiller-plough, with which, by going twice in the same place, with a man on the beam, we could cut a narrow furrow eighteen inches deep. A very short yoke was used on the oxen, so that the plough could be adjusted to cut only five or six inches in width. The ground was such that the plough would frequently encounter boulders as large as a man's head, and sometimes much larger. Many such stones would throw the plough out. But if the team could not draw them out with the plough, every one was dug out with a pick and crowbar. This was a slow way of preparing ground. But the process was thorough. As the work was performed in late autumn, when there was little else to be done with a team, we had ample time to put the plough down to a uniform depth over the entire field. Such preparation of the ground paid satisfactorily in the luxuriant and healthful growth of the trees.

The illustration given on page 60 (Fig. 36) represents a common plough, with an adjustable subsoil attachment for preparing ground for fruit-trees. One of the advantages of such a plough is, the best soil can be kept on the surface—where it always should be—and one man can hold an im-

plement that does the work of two ploughs
when held by two different men. In case a
light mucky soil were resting on a heavy
subsoil, it would be advisable to turn up
the compact subsoil, and mingle it with
the light soil by cross-ploughing several
times before the trees or apple-seeds were
planted. A few days' work with a double
team and an extra hand, when preparing
the ground for an orchard, will be labor
judiciously appropriated. If the ground is
in sod, and the soil is heavy, the preparation for trees should commence a year or

Fig. 36.

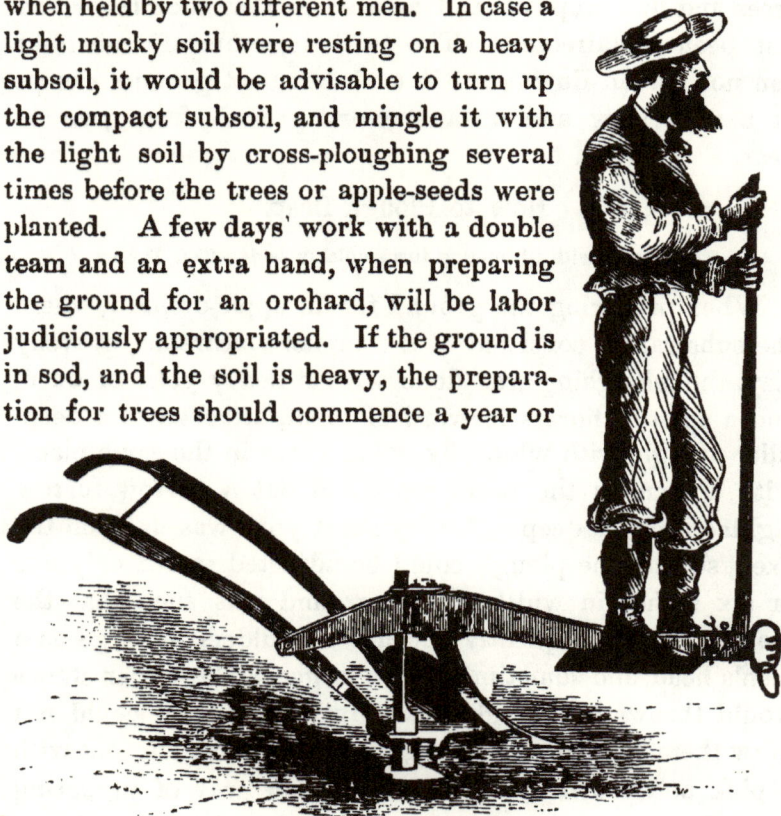

The true way to ride a plough-beam.

more before the seeds or trees are to be planted. It is folly
for a man to plant trees of any kind in grass ground. Grass,
clover, weeds, and grain, if allowed to grow near trees, will
retard their growth far more than one would suppose. As
a rule, this is true. And yet we have seen fruit-trees, on
deep alluvial soils, send out branches on every side, one to
two feet annually, even when the surface of the ground
around them was covered with a tough sod, and yielded a
heavy burden of grass.

Trench-ploughing.—Figure 37 represents a transverse section of trench-ploughing, as the ground will appear after having been ploughed by such an implement as is illustrated on page 57. If the land is not so

Fig. 37.

A section of trench-ploughing.

stony as to hinder the use of a trench-plough, by adjusting it to cut a furrow-slice about twelve or fourteen inches broad, the whole ground may be thoroughly pulverized to a depth of two feet. In order to perform the task well, the plough must cut narrow furrow-slices.

Under-draining Orchards.—Figure 38 furnishes a perspective of two rows of apple-trees, between which there is a deep tile-drain. When such drains are made, the joints of the tiles should be covered with collars, and the hardest

Fig. 38.

An under-drain between the rows.

earth should be placed on the tiles, rather than mellow soil, in order to prevent roots of trees from entering the joints, and enlarging to such an extent as to obstruct the water-passage. All under-drains between rows of fruit-trees should be three or four feet deep.

Drains between the Rows of Fruit-trees.—The ground for another orchard was prepared in the following manner: Stakes were set to indicate the rows of trees, thirty-three feet distant, each way. As the subsoil was a heavy, retentive calcareous clay, it was deemed necessary to put a deep under-drain between all the rows. The ground was ploughed by cutting very narrow furrow-slices, and going twice in a place, making a deep middle furrow midway between the rows. Then a good tile drain was put down, in no place less than thirty inches, and in many places four feet deep. The land was then ploughed again, turning back furrows to the ditches, and finishing deep dead furrows where the rows of trees were to stand. The tiles drew all the surplus water, so that the heavy and compact soil was greatly improved in its friability. But there are hundreds of orchards that would never be benefited by drains between the rows; and there are untold numbers of fruit-trees of all kinds which have never yielded half a crop of fruit, simply because the ground has been kept too wet and cold during two or more months in the former part of the growing season. There are frequently some portions of an orchard which are excessively wet, while most of the ground would never be benefited by under-draining. When young apple-trees do not grow luxuriantly where the soil is deep and fertile, and when the leaves do not seem to be healthy, it is a certain indication of one of the following difficulties, namely: The ground is too wet and cold; or borers are working in the stem; or the tap root has been cut off; or one was never sent down into the subsoil to bring up moisture in dry weather, and such inorganic material as the growing tree requires; or the soil round about the tree is not kept free from noxious weeds, injurious grass, and other crops.

If the rows are to be laid out in the *quincunx* style, stake them out so that the trees will stand in one direction, exactly parallel with the under-drains. By starting correctly, the rows may be made to run in any desired direction, either down a slope or across a slope, in a diagonal line. If tile-drains are made a rod from apple-trees, there will be very little danger that roots will ever fill the water-passage, provided the tiles are put down three feet. Were the tiles laid beneath or near the rows, roots would be liable to obstruct the water-passage in a few seasons. There need be no apprehension of rendering heavy land too dry by under-drains between the rows.

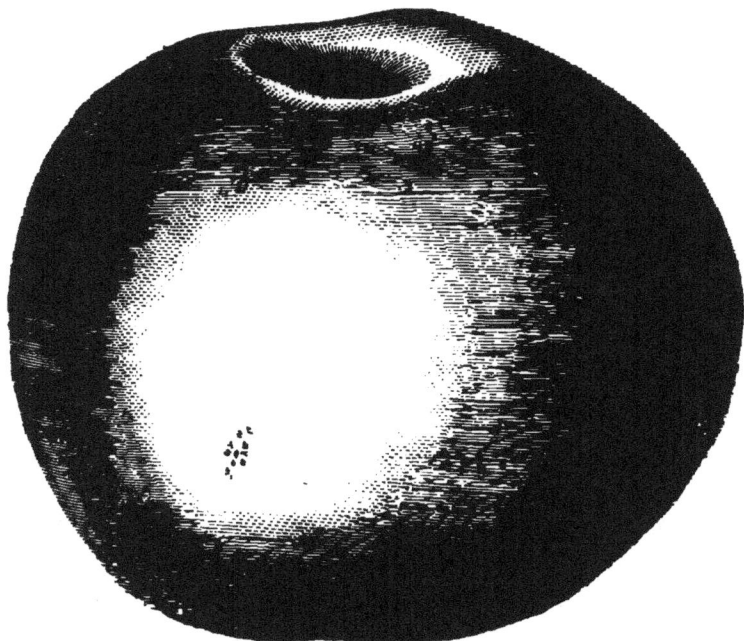

SWAAR.

The fruit of the Swaar is usually large, regularly formed, roundish, or roundish oblate; skin of a greenish-yellow when first gathered; but when fully ripe, it is of a fine golden color, dotted with numerous distinct brown specks, and sometimes faintly marbled with gray russet on the side and around the stalk; flesh yellowish, fine-grained and tender, with an exceedingly rich and aromatic flavor, and spicy smell. The core is small, and the skin tender. The trees are very productive on good land; season, from December to April.

The best Soil for Apples.—The most desirable soil for apples is one where there is a liberal supply of both clay and sand. Indeed, apple-trees *must* have both clay and sand, or they will not yield abundant crops of fruit for a long time. Consequently, a good fertile loam, which supplies both, will be found an excellent soil for an apple orchard. In numerous instances, a deep alluvial soil—if it happens to be of the right character—will be found excellent for either apples or pears. The fact that the roots of an apple-tree must be supplied with clay and sand, enables the beginner to improve his ground for apples, when either clay or sand is lacking, or when there is an excess of clay, and a deficiency of sand, and *vice versa.* Where the soil is deep and light, composed largely of sand and muck and leaf-mould, let two or three tons of clay be spread on the surface of the ground, in late autumn or winter, where each tree is to be planted. Then, in the spring, as it will be fine, let it be mingled with the light soil. On the contrary, if the soil is heavy, add sand, muck and leaf-mould; and mingle such additions thoroughly with the soil. Pile on, also, gas-house lime, old and quick-lime, leached, unleached, and coal ashes, chip-dirt, sawdust, fertile street dirt, scrapings of the manure-yard, tan bark, leather shavings, refuse of woollen mills, and all such manurial material as can readily be obtained. It will pay to cart sawdust two miles to put around apple-trees, as such material will furnish much of the best quality of food for the hungry roots of growing trees. Good barn-yard manure is also excellent for growing trees; and there is no danger of applying too much of it. But all such articles should be worked into the soil, where the roots can feed on such portions as will promote the growth of the trees and the development of the fruit.

We have met with scores of orchards in New Jersey

where the soil was deficient in clay; and after one or two fair crops had been gathered, the proprietors invariably complained that their trees did not yield even medium crops of fruit. But they did not seem to understand that apple-trees must have a supply of clay, as well as sand, in order to yield well. We have in mind large numbers of men who have fruit-trees on the light soils of New Jersey and of Long Island, who have always complained that their fruit-trees are failing, and that varieties are dying out. The fact is, there was nothing available in the soil out of which the roots of trees could produce apples. Consequently, those very men would report that in their localities "such and such varieties are complete failures;" when two tons of clay, a few bushels of lime, a top-dressing of ashes, and a dressing of rich compost, would have caused every branch to hang down with plump and luscious fruit. A gentleman of our acquaintance, at this suggestion, spread two wagon-loads of clay round about a barren fall-pippin tree, and the next season there was a bountiful crop of fine fruit. These suggestions will give a beginner a fair idea of the requirements of growing apple-trees, and of the chief operations in preparing the soil for a profitable orchard.

What growing Apple-trees require.—By referring to the analytical constituents of apples, it will be perceived that there is a large percentage of potash, phosphoric acid, some soda, some lime, and some phosphate of iron in the fruit. If these ingredients are scarce in the soil, we may succeed in producing fine-looking trees, but they will not bear fruit, except inferior crops of small, knotty, scabby, and one-sided specimens, simply because the roots can not find a supply of fruit-producing material in all the space they occupy. Bones, fish, dead animals, hair, leather shavings, and genuine superphosphate, and good marl, will supply very important ingredients to make large and luscious fruit. Ashes of all kinds—especially unleached wood-ashes—will supply potash, which is of eminent importance in the production of fair and plump fruit. Hence every gallon of soap-suds should be poured round about growing trees. During the production of all kinds of cereal grain, grass, vegetables, and fruit, Nature forms out of sand and potash a kind of liquid glass, which is spread over the leaves, stems, and fruit, to fortify the tender parts from the injurious effects of rust and mildew. Hence, if we top-dress growing wheat with a liberal supply of sand and wood-ashes, the stalks and leaves will be very harsh and stiff, and of a bright color, as Nature was supplied with an abundance of the right kind of material to form a glassy coat of mail over the surface of every part of the plant. Hence rust could not penetrate the glassy covering. Now, then, if we pile on the wood-ashes and sand, or soap-suds, an elastic, glassy covering will be spread so thickly over the surface of every apple, that rust and cracks on the fruit will seldom be met with. Were a person to cast a barrel or two of soft soap into the street, almost every thoughtful man would rebuke him for such a wanton waste of valuable fertilizing material for fruit-trees or for grain. And yet, after a barrel of soap has been dissolved and diluted in the wash-tub, it is quite as valuable for fertilizing purposes as before it was dissolved. There is very little danger of spreading too large a supply of wood-ashes around apple-trees. The roots must be fed with proper materials for developing the fruit. Three or four bushels of lime spread around a bearing tree, as far out as the branches extend, will often produce a fair crop of fruit. The carcass of a dead horse, or cow, if cut up into small pieces, and buried where the roots of an apple-tree will feed on the flesh and bones, will often make an indifferent bearer yield a bountiful crop. Marl is also excellent when used in the same way.

Requisite Elements of Fertility.—Whatever earthy ingre-dients are found in the wood, bark, and fruit of trees must be derived from the soil; and if the soil in which they are planted does not contain all these ingredients, the trees, or the fruit, or both, must necessarily fail. The mechanical condition of the soil is of as much importance in fruit-cul-ture as its chemical constitution. In its preparation, there-fore, the aim should be to secure good depth with perfect drainage. The object is to maintain a moderately luxuriant growth with early and thorough maturity in the wood; and this can be best attained in comparatively poor soils of mod-erate fertility. As a general rule, a soil which is dry, firm, mellow, and fertile, is well suited to the cultivation of fruit-trees. It should be deep, to allow the extension of the roots; dry, or else well drained, to prevent injury from stagnant water below the surface; firm, and not peaty or spongy, to preclude disaster from frost, mildew, and rust.

Most soils in this country may be much benefited for all decidedly hardy kinds of fruit, as the apple and pear, by good manuring. Shallow soils should be loosened deeply by heavy furrows and manure; or, if the whole surface can not be thus treated, a strip of ground eight feet wide, where the row of trees is to stand, should be rendered in this way deep and fertile for their growth. The manure should be thoroughly intermixed with the soil by repeated harrow-ings. The only trees which will not bear a high fertility are those brought originally from warmer countries, and liable to suffer from the frost of winter—as the peach, nec-tarine, and apricot; for they are stimulated to grow too late in the season, and frost strikes them when the wood is im-mature. It, however, happens, in the ordinary practice of the country, that where one peach or apricot tree is injured by too rich a cultivation, more than a hundred suffer by diminished growth from neglect.

One Error in manuring the Ground, when an orchard is growing, consists in applying too much carbonaceous material, when there is already too much, but a scarcity of mineral or inorganic material. Most soils contain a sufficient quantity of carbon, in the form of leaf-mould and muck, for the requirements of fruit-trees, provided there is no lack of inorganic elements in the soil, such as alumina, potash, and soda. Granite soils are among the best for fruit, as this rock abounds in feldspar and mica, both of which contain potash. As these rocks disintegrate and enter into the composition of the soil, they supply one of the most necessary elements for the formation of good trees and fruit. Some of the best orchards we have ever seen were on alluvial (loamy) soils, lying upon limestone rocks which came up near the surface. One of the best soils on which to raise fruit is that just cleared of a forest. The surface should be rolling or descending, and moderately dry and rich. Such ground needs little or no preparation. The roots of the forest-trees, as they decay, keep it loose and mellow, and afford the exact food necessary for a rapid and healthy growth of the fruit-trees; and in most instances, the soil abounds plentifully in those elements which are requisite to form the most perfect fruit. Another consideration, and a very important one, is, that fruit-trees grown on recently-cleared forest land are much less liable to be diseased than those grown on old land.

Any one going from an old settled country to a new one will not fail to observe the remarkable difference between the trees and fruit of the one and the other. How much more thrifty they are in the latter than in the former, and how much larger, fairer, and more perfect the fruit usually is on new land than on old.

Orchards on Rough Land.—There are thousands of acres of rough, broken, rocky, and uneven ground, in many States, which have never been ploughed, but which would be more valuable for orchards than for any other purpose, were apple-trees properly started on them. In many localities in New England, boulders cover the surface of the ground so completely that a team can scarcely be driven across the field. But the soil is generally of a superior quality for fruit of almost every kind, and especially for apples. Were the boulders removed, so that the ground could be ploughed, the land would yield abundant crops of cereal grain of almost any kind. But the boulders will hinder the growth of apple-trees but little, as such rocky soils are usually so porous and friable that roots will strike deeply and at a great length every season.

The most economical and expeditious way of starting an orchard on such ground is to commence, at one corner of the plot, to stake out the places for the trees. In many instances every boulder that would be in the way can be removed with crowbars and cant-hooks. Occasionally, however, a large boulder weighing ten or more tons may be lying exactly where a tree should stand. In some instances such large rocks can be buried in a hole dug close by, of sufficient depth to receive the entire stone beneath the surface. Whenever a rock is to be buried in such a manner, a strong brace should hold it from tumbling in before the hole is ready.

After the plot is properly staked out in quincunx order, drive a stake into the ground, a few inches from the points where the trees are to grow. Then make deep holes with a crowbar, as directed on another page; fill the holes with mould, sand, or mellow earth, and plant the apple-seeds in late autumn, as directed. Apple-trees will grow rapidly on such ground, and will yield abundant crops of fruit.

CHAPTER III.

LAYING OUT THE GROUND FOR AN ORCHARD.

"Now mark out the *quincunx* in hexagon lines,
And orderly stake it for trees or for vines;
If circles or triangles cover the ground,
At each intersection the *quincunx* is found."

MANY beginners are often at a great loss how to commence laying out an orchard correctly, so that the trees will stand in straight rows, at a right angle, or in the hexagonal or quincunx order. One can guess at a right angle, in some instances, with satisfactory accuracy; but sometimes a person may feel confident that his stakes have been stuck at a right angle, when he has found, to his disappointment, that the angle was surprisingly obtuse, or acute, as the case may be. If the rows are begun crooked, stake after stake may be altered without being able to form straight lines, and with only an increase of the confusion. If the first tree, in a row of fifty, be placed only six inches out of the way, and be followed as a guide for the rest, the last one will deviate fifty times six inches from a right line, even if the first error is not repeated. We have seen large apple-orchards with rows nearly as crooked as this. To say nothing of the deformed appearance to the eye, crooked rows prove exceedingly inconvenient every time the ground is planted and cultivated with crops in rows. It is a very easy task to mark out the ground for an orchard or vineyard, having the rows running at right angles, or in the quincunx order, without the aid of a surveyor's compass. The trees should stand in straight rows, first, for the sake of the workmanlike appearance; and, secondly, for the sake of convenience, when ploughing and cultivating the ground. When one sets out fruit-trees in crooked rows, the little crooks, which might have been avoided by a moment's fore-thought, often annoy through life, like mortifying mistakes in one's moral career, which can never be corrected. Hence the eminent importance of commencing correctly to mark out the ground, so that every row shall be as straight as can be desired.

The proper Distance apart.—Most persons when about to plant apple-trees, unless they have exercised themselves to think how much space an apple-tree, when full-grown, will occupy, are liable to make egregious mistakes by setting their trees decidedly too close in the rows. We have known many intelligent farmers set their young apple-trees only ten feet apart! After one tree had been planted, a distance of ten feet was measured from the tree, when an-other tree was set erect, and the eyes were glanced around to measure the range of the branches. The first impression was, that the tops would scarcely fill the apparently great area assigned to each tree. Hence trees were planted more than twice as close as they should be. The first apple-trees that we ever transplanted were set about twenty-seven feet apart. As we stood at our little tree—as boys are accustomed to, when setting out trees—and looked at those little whips, and then cast the eye about the large area that must be filled before the branches would touch each other, we thought, as a thousand other farmers do who have planted their trees too close together, that we should be an old man—were life prolonged—before the branches would meet each other. But see how soon

"Tall oaks from little acorns grow."

Seventeen years after the little whips were planted, we measured the height and breadth of the tops; and the height of one Benoni-tree, of fine proportions, was twenty-one and a half feet, and from the centre of the body to a point directly below the ends of the lateral branches, was fifteen to seventeen feet on all sides. If the branches of the other trees had spread out laterally with the same rapidity, their extremities would have interlocked a distance of four feet on all sides. Some pomologists insist that the distance between apple-trees should never be less than forty feet; while others say thirty feet; and others thirty-three feet; and some, even forty-eight feet apart. By a proper investigation of this subject, it will be seen that the proper distance will depend on the kind of trees to be transplanted. The correct idea is to plant the trees at such a distance apart that, when they have attained their full size, the lateral branches will not interweave each other. If trees be set so close together that the branches of one tree will grow among those on each side of it, the trees will not be as productive or healthy as if they were farther apart. Fruit-trees must be entirely isolated, in order to flourish and be productive. There should always be sufficient space between the tree-tops to permit a person, when gathering fruit, to carry a ladder in an erect position between the trees. If apple-trees stand two rods, or thirty-three feet, apart, in every direction, and if the trees are pruned correctly, from the first and second years of their growth, so that an area of fifteen feet on each side of every tree is properly filled with bearing branches, no one will ever have reason to complain that his trees were planted too closely. It would be a very easy thing to train some kinds of apple-trees to extend their lateral branches twenty feet each way, from the stem of the tree. But there would be no gain in such wide-spreading tree-tops. Most of our influential and experienced pomologists agree as to the distance of thirty-three feet for planting apple-trees, except for certain varie-

ties, like the Westfield Seek-no-further, Northern Spy, and
some others, the tops of which are never inclined to spread,
like the Tompkins County King. It is an excellent practice,
when planting an orchard, to plant varieties that grow erect,
alternately, with such trees as have wide-spreading tops.

Rows at Right Angles.—The diagram herewith given
(Fig. 39) will aid the intelligent
pomologist in staking out ground
so that the rows will run at right
angles, with as much accuracy as if
the angle had been formed with a
compass. The first step will be
to prepare as many stakes as there
are trees to be planted. Then set

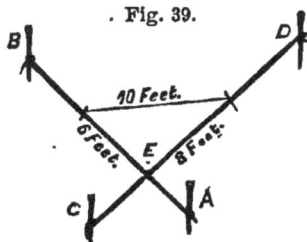

Fig. 39.

Laying out a right angle.

a row of stakes, say thirty-three feet apart, on the front
side of the orchard, represented by the line A B. The most
convenient way of measuring will be to make a light pole
of the desired length, so as to avoid all errors in measuring.
Now, to find the correct angle, stretch a chalk line, A B, in
the line of the first row of stakes. Then stretch another
line, C D, as represented, so that the lines will cross at E,
exactly at the desired corner of the plot. Measure with a
ten-foot pole on the line A B, six feet from the central intersection of the lines at E, and thrust a pin through the
line. At eight feet from E, stick another pin through the
line C D. If the lines cross each other at a right angle, the
measurement will be, as indicated by the diagram, six, eight,
and ten feet, from pins to the corner, and from pin to pin.
The next step will be to set a row of stakes in the line C
D. Fig. 40, page 74, will enable the operator to proceed in
setting the remainder of the stakes. Having set all the
stakes in the line A B, let them be set also in the line A D.
Then lay out the angle A B C, as shown by the diagram, and
set the stakes from B to C; after which, measure the dis-

4

Fig. 40.

```
*B  *  *  *  *  *  *C*
*   *  *  *  *  *  *
*   *  *  *  *  *  *
*   *  *  *  *  *  *
*   *  *  *  *  *  *
*   *  *  *  *  *  *
*A  *  *  *  *  *  D*
```

Rows at right angles.

tances from D to C, and set the stakes. A row of stakes now appears on every side of the orchard. If the soil is smooth and mellow, the most expeditious way will be to make a mark by dragging a heavy chain from one side of the orchard to the other; then set the stakes where the marks intersect. One person can stake out the ground in this manner with the most satisfactory exactness. After the stakes are all set, cast the eye along each row, and if any stake is one inch out of the line, let it be adjusted. Another good way, after setting the outside stakes, is to let one person set the remaining stakes, while two other persons—one on each side of the orchard—direct where to plant each stake, by sighting across the plot, from one outside stake to another. A ten-acre field can be staked out, according to the foregoing plan, in a few hours after the stakes are provided. A good substitute for one-half the stakes would be large corn-cobs stuck in the stake-holes, after the correct point has been found by employing tall stakes.

Planting Trees in the Quincunx Style.—The *ancient* quincunx style of planting trees consists in setting one tree at the central intersection of two lines crossing each other in a diagonal direction, from four trees set in the form of a hollow square. (See *Quincunx*, p. 329.) But trees can not be placed so uniformly over the ground when planted in this manner as when set out in the usual way, with rows extending in two directions. The *modern* quincunx style of planting trees consists in setting them in such a manner that the distance from one to another, throughout the orchard or vineyard, shall be uniform. In other words, any

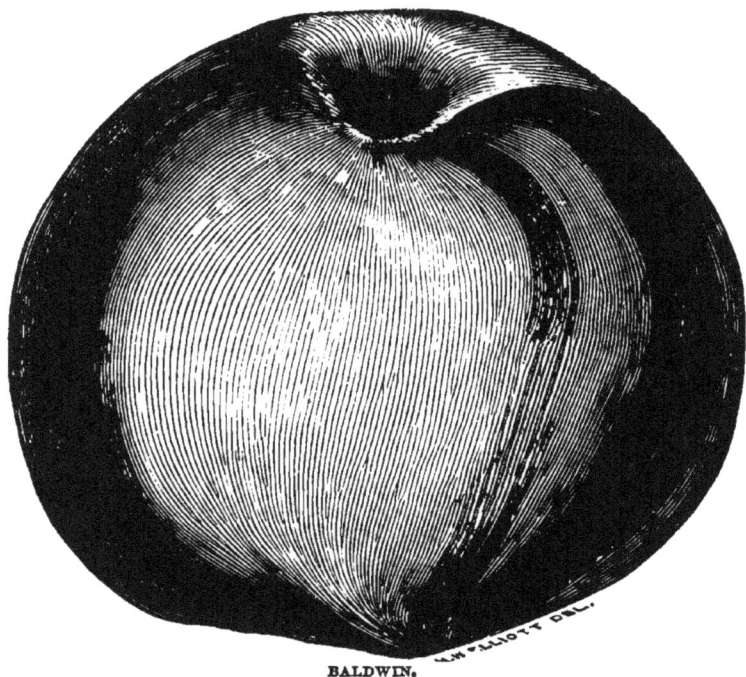

BALDWIN.

Synonyms.—Steele's Red Winter, Butters, Red Baldwin, Pecker, Woodpecker, and Felch. The fruit is of good size, roundish, striped with crimson; in the sunshine, the apples are often of an orange and red color, covered with a few russet dots, and with radiating streaks of russet about the stem; flesh, yellowish-white, crisp, with an agreeable mingling of the saccharine and acid, which constitutes a rich flavor. The tree is a fine bearer, vigorous and upright in growth; season, from November to March. The Baldwin is one of the most profitable apples raised for market.

four trees will be found to stand at the angles of a plot of ground of a rhomboidal form, and any tree in the orchard within the outside row will stand exactly in the centre of a circle, in which circle six other trees are standing all at a uniform distance apart. If apple-trees, for example, be set thirty feet apart, in the usual manner, or in the *old* quincunx order, one row more of trees may be set on one acre by adopting the *modern* quincunx style; and yet all the trees will be thirty feet distant from each other over the entire orchard. Planting fruit-trees or grape-vines in the

modern quincunx order, distributes them more uniformly
over the ground than if planted in any other manner.
Whatever is put out in this style will stand in straight
rows in four different directions.   Indian corn, or any
other crop that grows in hills, may be cultivated in the
quincunx order more satisfactorily and profitably than if
the rows were to cross each other at right angles.   When
rows are formed in the modern quincunx order, the horse-
hoe may be run four different ways between them, thus en-
abling the cultivator or horse-hoe to do the work of pulverization, or rooting up the weeds and grass, more thoroughly than when the implement can be driven only one or two
ways.

Planting trees and grape-vines in the modern quincunx
order has frequently been recommended as the most desirable way of putting them out.   But the writers told every
thing about the operation except what a practical man desires to know, and what a beginner *must* understand before
he can proceed to lay out the ground correctly.   Some writers—who probably never attempted to stake out ground
in the quincunx order—have even given the number of feet,
inches, and fractions of an inch, from tree to tree, when
measuring at a right angle.   But working-men, who are not
accustomed to measuring inches and fractions of inches,
need a more convenient and expeditious way of staking out
ground than to measure with a two-foot rule, from one
point to another, on the surface.   When the ground is correctly laid out in the modern quincunx order, the plot between any three corners, or between any three trees, will be
exactly in the form of an equilateral triangle.   If, for example, it were desirable to mark out the ground in a large field
for planting Indian corn or potatoes in the modern quincunx
order, the correct and easy way would be to do the marking as usual, all one way, first, by commencing on the long-

est side of the field. Then, after the marks have all been made in one direction, the cross-marking should be commenced near the middle of the plot, or where the rows will be the longest, marking each way from the middle of the field. The idea will be to run the marker across the first marks at an angle of about sixty degrees. The correct direction may be readily ascertained by the aid of a surveyor's compass. But an inexperienced beginner seldom has the advantage of such an instrument; and even if he does possess one, there are thousands who must rely on some more simple and cheap device to aid them in the performance of such a job.

The accompanying illustration, representing a *sweep* for marking out ground, is a figure of a device which we have employed in laying out the ground for grapevines and fruit-trees. The sweep was made by securing two light strips of wood, A A, to the stakes B B, with small round

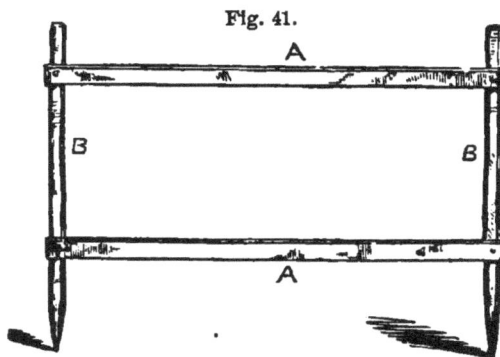

Fig. 41.

A sweep for marking out the places for trees.

bolts, or with large wood screws, so that the horizontal bars may be turned a little, like a hinge. The stakes should be about five feet long, two inches in diameter, either round or square, and pointed, so that the lower ends may be thrust into the ground at pleasure. With such a device, two boys, or illiterate laborers, who do not know an angle from a parallel of latitude, may proceed to mark out ground for any thing with as much facility and dispatch as an experienced surveyor can do it with a compass. It will be very difficult to do the work wrong if the first row is only staked out straight. It will be understood that the stakes of the sweep must be set as one would adjust the points of dividers, or the points of a mechanic's compass, just as far apart as the trees are to be planted. After the sweep is ready for use, the following diagram will aid still further in finding the exact spot where each tree or vine is to stand.

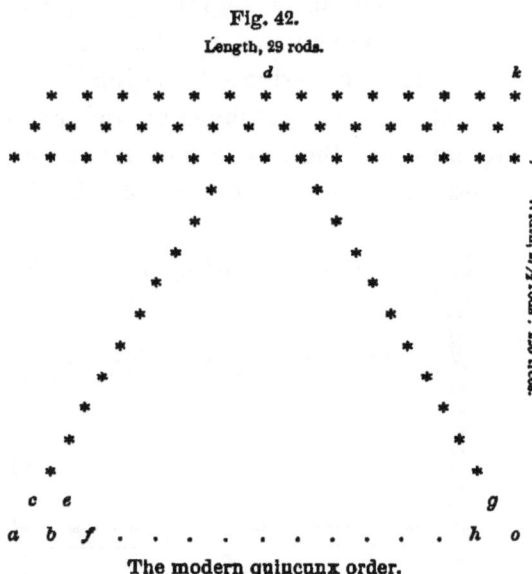

Fig. 42.

Length, 29 rods.

(diagram of quincunx planting pattern with points labeled d, k at top and a, b, f, c, e, g, h, o at bottom; text on right margin reads "Width, 27½ rods : 225 trees.")

The modern quincunx order.

We will suppose, for example, that a field of a square or irregular form is to be planted with apple-trees to stand thirty feet apart. Commence on the longest side of the field, and stick one row of stakes in a straight line, exactly thirty feet apart. The distances should be indicated with the points of the sweep. See that this row of stakes stands in a straight line. If there be any crooks in the first row, the same trifling variation will be multiplied as the marking progresses. Let the stakes be driven at least one foot into the ground, so that the point where the tree is to stand may not be lost. Now, to ascertain the places for the trees in the second row, take up the end stake in the first row, and plant one point of the sweep in the stake-hole at *a* (Fig. 42). There should be one person at each end of the sweep. The point of the sweep at *a* should be held securely in the hole while the person at *b* describes a part of a circle with his point of the sweep, about where he judges the tree must stand. Now let *a* carry his end of the sweep to *c*, and plant the point in the stake-hole, and hold the sweep firmly while the operator at *b* makes a mark with his point of the sweep across the first mark. Here, then, will be three points indicated with as much mathematical precision as if they had been found by the aid of a surveyor's compass. The beginning is practically as perfect as any piece of work can be; and if the operators will be careful to plant every stake exactly at the central intersection of the marks made by the point of the sweep, as at *a, b, c,* the stakes will all stand in rows satisfactorily straight when the work of laying out the ground has been finished. Let the person at *b* thrust the point of the sweep into the ground, and set a stake in the hole. Then carry the end of the sweep to *e,* and make a circular mark, say four feet in length. Now let one of the points of the sweep be placed in the stake-hole at *f,* and the place where the circular marks cross each other at *e* will be the

Fig. 43.

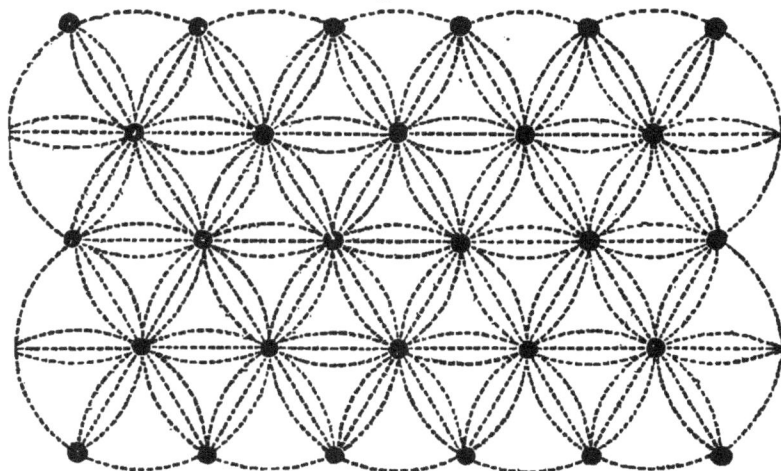

Orchard laid out in the modern quincunx order.

point to make a hole for the stake at *e*. This explanation will be sufficient to enable any person of ordinary intelligence to stake out his ground with as much dispatch as the stakes can be set in the usual manner. After the stakes have all been put in their places, should some of them vary from a line, they may readily be set straight. If the ground be all marked out in circles just touching each other, the points at which the circles intersect will show the places to plant trees.

Staking out Ground in Triangles.—The ground can be staked out in triangles with no other measure than a ten or twenty foot pole, with as much accuracy as it can be done with a surveyor's compass. We will suppose, for example, that the trees are to be set thirty-three feet apart. In the first place, make a light pole of the desired length, thirty-three feet, and set a row of stakes on the longest side—if there be any difference in the distance—of the ground, as represented by *a o*, in the diagram, page 78. Let the stakes be small and straight, and at least four feet high. Now let one person hold one end of the pole at the end stake, *a*, while another makes a mark at the other end, *c*. Then let the end at *a* be carried to *b*, while the end at *c* is retained at the same point where the two marks at *c* intersect. Let a small stake be set at *c*. Next place the pole from *b* to *e*, and from *e* to *f*. At *e* and *c* will be the places to set trees. Let the stakes at *a c* be set perpendicularly. Now let a stake be set up at *d*, so that *a c* and *d* will stand in a line. Then set a stake every thirty-three feet—or the length of the measuring-pole apart—in the line from *c* to *d*. Let the measuring be done with accuracy, and let the stakes be set perpendicularly. This second row of stakes will constitute a reliable guide to aid in finding the correct points for other stakes. Now with the measuring-pole find the point for another stake at *g*, by placing one end of the pole at *o*, and the other end at g. Having set a stake at o, and one at g, measure with the pole, and set stakes in the line $o \ g \ d$, the length of the pole apart. The principal points of the entire plot are now fixed with such accuracy that one person may stand at h, and move towards a, while another sets all the stakes in the row, $o \ k$. After a row has been set from a to o, a row across the ends at $d \ k$, another row from a to d, and from o to d, the remainder of the stakes can be stuck at the desired points, without measuring, by simply setting each stake in a line, in two directions, with those stakes that have been driven in the right places.

It will be perceived by the two diagrams given on pages 74 and 78, that they both represent a plot of ground twenty-nine rods long by twenty-seven and a half wide, embracing just ten square rods over five acres. The spaces indicated between the places for trees, as may be seen, represent exactly two rods, or thirty-three feet. The points may be counted in the diagrams quite as well as if a five-acre lot were before us all properly staked out. If trees were to be planted on two such plots, there would be a gain of space for thirty trees by planting in the quincunx order, while no two or three would be crowded. If the trees were to be set only half the distance apart, there would be a gain of space for sixty trees, as any one can calculate for himself.

In case it were desirable to set pear-trees, peach-trees, or grape-vines in the quincunx order, all that will be requisite is to make a pole of the desired length, and set a row of stakes six, ten, or sixteen feet apart, on one side of the plot at $a \ o$, Fig. 42. Then proceed to measure with the pole as directed, to find the proper angle for setting the other rows.

Another way for marking out ground in a triangular form, or in the quincunx order, is to procure three poles of a uniform length—say sixteen and a half feet long—and bolt or

Fig. 44.

A triangular rule for laying out orchards.

nail the corners together, as represented by Fig. 44, which will give the same results as laying out by circles, or as directed in the preceding paragraphs. It will be seen that three poles of equal length, placed with the ends together as represented, will give the exact angle for a hexagonal figure, for an equilateral triangle, and for the modern quincunx order. Let a piece of plank be nailed in each corner as shown, and a two-inch hole bored at the end of each piece for stakes, and a rule will be produced for laying out ground for trees which any person can use without making mistakes. After a person has used such a rule for a few minutes, he will be pleased with the great convenience of the arrangement. Three straight and smooth stakes will be required—one for each hole or corner of the rule.

We once marked out the ground for grape-vines, in this manner, eight feet apart; and also the ground for peach-trees, sixteen feet apart, by the assistance of a boy, who quickly learned to set his stakes with satisfactory accuracy. If the operators are careful to measure accurately, there will be no more difficulty in planting trees in the quincunx order than in rows running at right angles.

To facilitate the labor of getting every tree in the desired position, procure a board, *b* (Fig. 45, represented on the following page), about ten feet long and eight inches wide, and scollop out a half-circle for the stake, *a*, which stake represents the exact point where the tree is to stand. When the hole is about to be dug, where the stake, *a*, stands, place the board to the stake, as shown; then put two smooth stakes, *c, c*, through holes, into the ground. The stakes, *c, c,* must Fig. 45. remain until the tree is in its place. Remove now the stake, *a,* and board, *b,* dig the hole, put the board, *b,* back over the stakes *c, c,* and set the tree in the scollop where the stake *a* stood before the hole was dug. By using such fixtures, one person can plant trees with the most satisfactory accuracy,

A gauge-board.

without having a second or third helper to hold the tree. If the stake *a* is in the correct spot, the tree will stand in line.

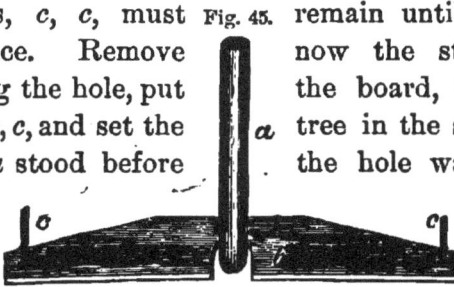

TRANSCENDENT CRAB-APPLE.

Fruit, medium to large for its class; roundish-oblong, flattened at its ends, slightly but regularly ribbed; golden yellow, with a rich, crimson red cheek in the sun, covered with a delicate white bloom; when fully ripe, the red nearly cov-, ers the whole surface. Stem, long and slender, set in an open, deep cavity; calyx closed, with long reflexed segments; flesh, creamy-yellow, crisp, sub-acid, a little astringent until fully mellow, when it is pleasant and agreeable; seeds, full and abundant; leaf, broad, oval, with an acute point, and narrow, sharp, regular serratures; season, early autumn to winter. The stem is represented much shorter than it usually appears on most of the apples.

CHAPTER IV.

REMOVING AND TRANSPLANTING BOTH OLD AND YOUNG TREES.

The crowded roots demand enlargement now,
And transplantation in an ampler space.
Indulged in what they wish, they soon supply
Large foliage, o'ershadowing golden flowers,
Blown on the summit of the promised fruit.—COWPER's *Task*.

THE chief idea to be kept in mind, in order to transplant any tree with satisfactory success, is to take up every root and rootlet without the least mutilation, and bury them all back again in a mellow seed-bed, no deeper nor more shallow than they were when taken up. If this could be done properly, no tree would ever suffer injury from a removal. The act of planting a tree is an important one. We commit it to the soil, as we would send a youth into the world, to sustain a separate and independent existence. Nursed no longer under the eye of the propagator, it must contend with the storms, floods, and vicissitudes of climate, with a broiling sun and killing frost, developing as it best may its system of branches, buds, leaves, blossoms, and fruit, according to its nature, for our especial benefit, as well as to add to the beauty of this pleasing world. How scrupulously careful, then, should we be in planting a tree—an apple-tree, perchance, which we expect to pour an annual offering of rich, glowing, and luscious fruit in our lap; or a pear, whose melting gifts are eagerly sought after by impatient cultivators, too willing to reap the harvest before they have sown the crop, with intelligence, skill, and perseverance; or, still more pleasing, the soft and roseate peach, so often nipped in the bud by stern winter's frost. There are few operations in our husbandry in which so much want of reflection, not to say gross and willful neglect, is displayed as in the transplanting of trees, whether for fruit or ornament. It must, however, be admitted that in this, as in all other branches of rural industry, great improvement has taken place within the last thirty or forty years, since agriculturists commenced to read and reason on the principles that are involved and brought into action and practice of every branch of their business, as well as in the manufacturing of leather, iron, or any other article. Roots of trees usually strike downward instead of upward; and nearly all kinds of trees, whether fruit or forest trees, send out a system of roots just below the surface of the ground. If we examine the forest-trees—unless they are standing on a very loose and porous soil—we find that, although the *tap*-roots of trees may strike several feet deep in the earth, there is a system of numerous roots near the surface of the earth, on which the trees depend almost entirely for sustenance. Let earth be hauled and placed around a flourishing tree, a foot or more in depth, and that tree will soon send out a system of roots all around it near the *surface* of the soil. It is absolutely essential that all the interstices among the roots be well filled, settling the fine earth, if need be, by pouring in water. If large cavities are left the frost may destroy them. This is the reason that some persons have been unsuccessful in keeping trees through winter.

Suggestions to be observed when Transplanting Trees.— Beginners in fruit-culture will do well to read over the following concise rules, until they thoroughly understand the correct theory and practice of transplanting any kind of trees or vines :

1. When a tree is taken up, endeavor to take at least one-twentieth part of the small roots. The practice usually is

to leave so many of the roots where a tree grew, that it is no exaggeration to say that the roots were left behind.

2. Trees should always be set about as deep as they stood before digging up.

3. A small or moderate-sized tree at the time of transplanting, will usually be a large bearing tree sooner than a larger tree set out at the same time, and which is checked necessarily in growth by removal.

KING OF TOMPKINS COUNTY.

Synonyms.—King Apple, Tom's Red, Tommy Red. The fruit of this apple is very large and globular, inclining to conic, sometimes oblate and angular: the ground color is usually of a yellowish hue, shaded with red, striped and splashed with crimson; the flesh is yellowish, sometimes rather coarse, but juicy and tender, with an exceedingly agreeable and rich vinous flavor, delightfully aromatic. This is one of the choicest varieties for almost any locality; season, from December to March. The tree has a large and spreading top, and produces abundantly.

4. Manure should never be placed in contact with the roots of a tree in setting it out; but finely pulverized earthy compost may be employed. Young trees may be manured to great advantage by spreading manure over a circle whose radius is equal to the height of the tree, in autumn or early winter, and spading this manure in, in spring.

5. If the roots of a tree are frozen out of the ground, and thawed again while in contact with air, the tree will be killed. If the frozen roots are well buried, filling all cavities before thawing any at all, the tree will be uninjured.

6. Never set young trees in a grass field, or among wheat or other sowed grain. Clover is still worse, as the roots go deep, and rob the tree roots. The whole surface should be clean and mellow; or, if any crops are suffered, they should be potatoes, carrots, turnips, or other low hoed crops. If an area round about each tree, as far as the branches extend, be properly mulched with coarse manure, grass or clover may be cultivated in an orchard.

7. The roots extend nearly as far on each side as the height of the tree; hence, to dig up the soil by cutting a circle with the spade, two feet in diameter, the spade will sever nine-tenths of the roots; and to spade a little circle about a young tree, not one-quarter as far as the roots extend, and call it "cultivation," is like Falstaff's men claiming spurs and a shirt collar for a complete suit of clothes.

8. Watering a tree in dry weather affords but temporary relief, and often does more harm than good by crusting the surface. Keeping the surface constantly mellow is much more valuable and important; or if this can not be done, mulch well. If watering is ever done from necessity, remove the top earth, pour in the water, and then replace the earth; then mulch, or keep the surface very mellow. Shrivelled trees may be made plump before planting, by covering tops and all with earth for several days. Water-

ing trees before they expand their leaves should not be done by pouring water at the roots, but by keeping the *bark* of the stem and branches frequently or constantly moist. Trees in leaf and in rapid growth may be watered at the roots, if done properly, as directed.

9. Warm valleys with a rich soil are more liable to cause destruction to trees or their crops by cold than moderate hills of more exposure, and with less fertile soil, the cold air settling at the bottom of the valleys during the sharpest frosts, and the rich soil making the trees grow too late in autumn, without ripening and hardening the wood of the branches.

10. Drain the ground if it is at all wet; and commence the work of pulverization and subsoiling at least one year before young trees or seeds are planted.

11. Remember that fruit-trees need to be fed every year, in order that they may yield large, smooth, and luscious fruit. Trees can not concoct fruit out of material that will not make fruit.

12. Recollect that it is far better to procure trees from a poor soil than from one that has been highly manured; and it is better still to rear your own trees from the seed where they are to grow.

13. The excavations where trees are to stand should be so broad that the roots can readily spread to the next row without meeting with unbroken ridges of hard-pan.

14. Beware of planting fruit-trees too deep, especially on heavy soils. On light, loamy soils they should be set, in some instances, four to six inches deeper than on a heavy soil, according to the lightness and porosity of the land. It is a safe rule, however, to set no deeper than the trees stood in the nursery; and this can easily be determined by their appearance at the base. Every fibre should be extended in its proper direction, and carefully surrounded with compost.

No cavities should be left in covering the roots, nor should one be injured by the hand or spade.

15. Let every tree be transplanted as soon as practicable after it is taken up, and let the roots always be protected from sunshine and drying winds until they can be buried in the soil. Drying and exposure to the air always injures roots. The longer the exposure and the greater the drying process, the greater, of course, is the injury. Digging up trees when destitute of leaves, and leaving them an hour or two in the shade, produces little or no harm; but they should not be allowed to remain in the sun, nor be exposed for a whole day to the wind. If they can not be set out or packed immediately, they should have the roots plunged in a bed of mud, to give the surface a thin coating; or the roots should be immediately buried in mellow soil or sand, until further operations are commenced with them.

16. When trees of any kind are purchased, it is always better, in every respect, to choose young trees, say two or three years old, that are vigorous and bushy, than to purchase large ones, four or five years old. As a rule, fruit-trees that are far brought are dearly bought; and a person seldom orders trees from a distance but once. By looking around, most persons can find good trees in their own town. But the safest and best way is to raise the trees where they are to grow. It does not require so long a time to rear an orchard as many seem to think.

Practical Operations.—The illustration herewith given (Fig. 46, p. 90) will give the beginner a more definite idea of the great extent of the roots of a young tree, and of the labor required in transplanting a fruit correctly. After the excavation has been made sufficiently broad and deep to receive every root when extended to its full length, place the gauge-board, *b* (Fig. 45), over the pins, *c, c,* and set the tree with the body in the scollop, holding it erect with one hand while a few

Fig. 46.

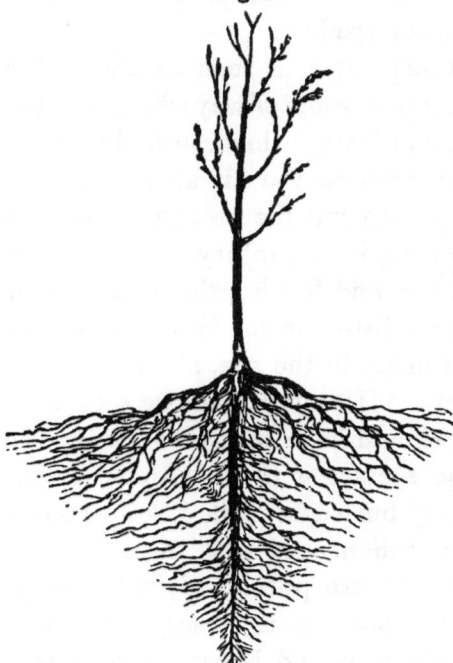

Extent of the roots of a young apple-tree.

shovelfuls of mellow soil are scattered over the roots. If the ground is at all compact in the bot-tom of the excavation, make a deep hole with a large crowbar, to receive the tap-root, if there is such an appendage; and if a long lateral can be obtained, let it be thrust down into the hole, af-ter which fill the hole with mellow soil or rich mould. The object of the tap-root is to supply the tree with moisture in hot and dry weather. If the ground is poor, let a load of mellow soil be carted from a distant field, or plough up the turf along the highway for the purpose. Spread out the little fibres, and cover them with rich, mellow soil. Much of the success of transplanting depends on bringing the soil in contact with every fibre, so as to leave no hollow to cause the decay of the roots. Not only must this be secured by patiently fill-ing in the cavities among and *under* the roots, but, where the trees are not small, it is customary to pour in a pail of water when the roots are nearly covered with soil. The water carries the liquid mould to every hidden part. After the water has settled away, fill up the hole. Avoid press-ing the earth about the tree with the foot. In windy situations it will be necessary to place a stake each side of the tree, to hold it upright; or, if large stones are at hand, place a layer of them near the tree, which will not only secure the tree from moving to and fro, but will often prevent teams and implements from passing over small trees.

Before any tree is transplanted, the end of every mutilated root should be cut off smoothly. If roots have been bruised badly, it is always better to cut off every thing beyond the bruise.

Staking Fruit-trees.—The accompanying illustration (Fig. 47) represents the correct way of staking trees. When a stake is driven down near a small tree, it often obstructs the growth of the roots; and if a large stake is driven after a tree is transplanted, a root is sometimes badly mutilated. But if stakes are set as represented by the figure, they will not interfere with the roots. Fruit-trees should never be staked for the purpose of holding a tall and slender tree erect. When a young tree has run up so slim that the wind often bends the top half way to the ground, the true way is to cut the central stem off, say a foot or more, for the purpose of inducing the tree to grow more stocky and bushy. (See *Pruning*.) When a hundred or

Fig. 47.

Staking a small tree.

more of such stakes are needed, let a log of some durable timber be sawed into stuff, say two inches square. Then let the pieces be thoroughly seasoned before they are driven into the ground, and let the lower ends be well smeared with coal-tar. A hole should be made with a crowbar, the stake driven in firmly, the upper end sawed off in a horizontal direction, and a piece of board nailed to the top of the stake. Each board should have a hole bored through it near one end; and when it is to be put on, let it be split in two through the hole, and the pieces nailed to the top of the stake. Pieces

of canvas should be drawn into the hole, to prevent injury
to the bark as the tree is blown to and fro by the wind.   In
some instances young trees are not rooted with sufficient
firmness to prevent the stem from working a large hole in
the ground at the collar of the tree.   When the ground is
frozen, if such trees are not held by a stake, the bark is often
worn entirely through to the wood by the incessant swaying
of the tree.

Excavating with a Scraper.—One of the most satisfactory
modes of excavating for fruit-trees that we have ever practised is to stake out the ground, and then plough a small plot,
say eight to ten feet wide by fifteen feet long, hitch the
team to a good dirt-scraper (Fig. 48), and scrape the loos-

Fig. 48.

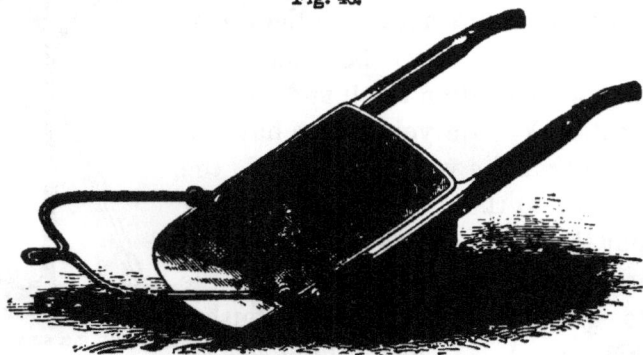

A two-horse cast-iron scraper.

ened earth each way from the point where the stake stood.
After all the loosened earth is scraped out, hitch again to
the plough.   It may be necessary to employ a piece of log-
chain one or two feet long between the plough-clevis and
whiffle-trees, with a man on the beam if the ground is very
compact.   Let the wheel beneath the beam of the plough be
adjusted so that the plough may enter only four or five inches.
Then let one person "ride the plough-beam" (see the illustration of a man riding a plough-beam, p. 60), and the team

will be able to draw the plough through the heavy and solid earth. If no gauge-wheel is employed, the plough will often plunge in so deep that the team can not pull it through. Excavations can be made with plough and scraper two feet deep in a very economical manner, when one has a good team. After the earth has been removed to the desired depth, let it be returned with the scraper, until the depressions are sufficiently full to receive the trees. Then set up a stake where the tree is to be planted, and proceed to excavate for another tree.

We once tested this mode of preparing holes for fruit-trees, and found that a man and boy, with a span of horses, could excavate a hollow about two feet deep, and return the earth in about one hour per hole. But it paid well in the growth of the trees.

Setting Trees on the Surface of the Ground. — As has previously been stated, there is danger of planting trees too deep, as many of the roots strike downward into the earth, as naturally as the tops stretch upward into the air, when there is not some impenetrable obstruction to hinder their ramification through the soil. None except the annual feeders near the surface of the ground are found to tend upward, as they push out from a main root, or branch of a root. In case a tree is planted too deep, a new system of roots will be thrown out from the stem, one or two inches beneath the surface of the soil. No matter how deeply tap-roots and branch-roots may be sent downward, it is a *habit* of fruit-trees to produce a complete system of roots, rootlets, fibres, and feeders so near the surface of the earth, that the countless number of minute mouths may imbibe the moisture and plant-food soon after the small particles begin to descend from the surface.

A knowledge of these facts once warranted an experiment in planting trees directly on the surface of the ground.

In the spring of 1843, when planting the first trees for our own orchard, there were a few places where holes could not be dug with a spade, as the earth for several feet in depth was composed of fragments of slate and small boulders. Neither could the ground be ploughed. A thick and tough sod of Kentucky blue-grass rested on the surface; but roots of trees could spread among the stones without difficulty. Hence we resolved to try the doubtful experiment of setting trees without digging any holes. Stake-holes were first worked down through the stones with a crow-bar, and stakes were driven in firmly to support the trees. The roots were then spread out on the grassy sod as a tree was held near a stake, after which a few bushels of mellow soil were shovelled from a loaded wagon around each tree, sufficient to cover the roots with about two inches of mellow dirt. The body of each tree was then tied firmly to the stake; and the surface of the ground round about the trees was mulched with coarse, strawy barn-yard manure, covering an area of about eight feet in diameter. Pieces of boards, old rails, and brush were laid on the mulch to prevent fowls from removing the coarse material. Amidst the sneers of those who knew that such a mode of planting was superlatively ridiculous, and could never prove at all satisfactory, we waited in doubtful suspense for the result. The trees that were set in mellow ground, around which the surface was kept clean and free from vegetation, grew a few inches higher and broader, while every tree that was set on the grassy surface threw out branches of good ripe wood, laterally and vertically, from one to four feet in length, before winter. The first season some gentlemen called to learn the secret of such a wonderful growth, and measured the new wood just before the leaves had fallen, and found many branches over four feet in length. After the first season, the branches did not grow faster than those on other trees. Every tree that was planted on the surface succeeded in a most satisfactory manner, and yielded as much fruit as any others. In 1868 we saw them bending beneath a bountiful burden of fruit. In after years we planted many trees on the grassy surface, always with the most satisfactory results. On stony fields, in rocky dells, and on lawns, where the ground can not be ploughed, and where disturbing the sod is objectionable, there is no more satisfactory way of planting trees of any kind, or bushes, than to spread out the roots on the sod and cover them with a little mellow earth; then mulch the surface.

Effects of Planting too Deep.—The roots of trees are things of life; and they must have the benefit of the air as well as the growing branches. Almost any fruit-tree may be killed in one or two seasons—except when it stands in certain favored spots—simply by filling earth around it, say two or more feet deep. Many a valuable tree, standing in a low place, has been killed in consequence of the earth being filled in around it when grading the surface of the ground.

We well remember a large apple-tree, now in full bearing, which was taken up in the winter, with a large ball of frozen earth at the roots, and placed in a deep hole in the orchard. It was probably set too deep; because, for nearly fifteen years, that tree did not flourish well nor produce fruit even *tolerably* fair. But as soon as it sent out a system of roots near the surface of the ground, it began to grow rapidly, and to bear abundantly, and is, at the present writing, a flourishing and productive tree.

The first apple-trees that we transplanted were set too deep, as we found, three years afterwards, that they were sending out a system of roots all around the bodies, near the *surface* of the ground, while about four inches *below*

PECK'S PLEASANT.

Synonym.—Waltz Apple. The fruit is rather above medium size, roundish, a little ribbed, and in most instances slightly flattened, with an indistinct furrow on one side. The skin is smooth, and green when first gathered, often covered with a little dark red. When fully ripe, the fruit is covered with a beautiful clear yellow, having a bright blush on the sunny side near the stem, where the surface is marked with gray dots. The flesh is yellowish and fine-grained, juicy, crisp, and tender, with a delicious, highly aromatic, sprightly sub-acid taste; quality, good to best; season, November to March. This apple passes for an excellent apple both for market and for culinary purposes.

these were the main roots; and as soon as these new roots had attained a suitable size the trees began to flourish luxuriantly.

The mice gnawed one of our apple-trees, more than two-thirds of the distance around it, for two feet above the surface of the ground. We raised a mound of mellow earth around it, a few inches higher than it had been gnawed, and covered it with sods. The next season the hens scratched away some of this earth, and we discovered that there was a system of young roots from six to ten inches long just below the *surface* of this mound. From these considerations, it is safe to conclude that there is great danger of planting trees *too deep*, and that the *most proper* depth is not more than two or three inches below the surface of the ground. We always fill the holes for trees even-full with the general surface of the ground, before the trees are placed in their position; and we have always found that they would settle down in a few months as far as they should ever be set. If the soil is well pulverized beneath the roots, they will soon occupy the whole space below, and will take a firmer hold of the earth than if they were placed as deep in the mellow-bed as they stood in the nursery. Allowance must be made for settling, when the soil is very wet.

Transplanting in Autumn, or in the Spring.—The question is, which season is the better one? In some instances autumn is preferable, and in others spring offers advantages. If the condition of soil and trees are the same, we have never been able to perceive any difference. Still both periods have their advocates; and there are about as many arguments in favor of one season as the other. If the work be well performed in the spring, just before the buds begin to swell, trees can not be transplanted in a more favorable period than that. When apple-trees are set out in spring, they come fresh and vigorous from the ground, from which they are withdrawn easily, the soil having been mellowed by the winter frost. As soon as out, they are ready to recover by immediate growth from the effects of removal; the buds swell and expand into fresh leaves, and new shoots rapidly

5

spring upward. More than this, if trees are taken up in the spring, before the buds expand, and are transplanted at once, the mutilated roots will heal and send out new rootlets sooner than if they had been planted in autumn. Our long experience is decidedly in favor of spring planting for every thing that must be transplanted. But a far better way is to plant the seed where trees are to grow, and thus avoid transplanting. When trees are planted in autumn the roots take no hold of the ground until the next spring. Hence a crowning and unanswerable argument in favor of planting in the spring of the year is, that it is far better for trees or vines of any kind to have a vital connection with the soil during the drying and cold weather, than for all vital connection to be severed for from four to six months. This fact alone is sufficient to show that spring planting is preferable to autumn. But if trees can not be properly managed in the spring, and if they can not be planted till late in the growing season, it is better to plant in autumn. We have observed that during several years past many Western and North-western farmers have reported the loss of thousands of apple-trees, all in consequence of being planted in the spring.

Double - planting Orchards.—The practice with some pomologists, when they first commenced planting apple-orchards, was to set a row of peach-trees alternately between the apple, in two directions, by which they were enabled to plant out twice the number of peach-trees on a given area as there were apple-trees. Then, by the time the apple-trees had attained sufficient size to require the space occupied by the peach-trees, most of the peach-trees would have reached the limit of their existence, so that they could be removed. The results were eminently satisfactory.

The question is frequently asked, at the present day, whether dwarf and standard pears, peaches, plums, or other fruits could be planted in the intervals of the apple-trees without injury to the latter? In most instances, with fair cultivation, on a fertile soil, apple-trees, thirty-three feet distant, will in the course of some sixteen to twenty years nearly cover the whole ground, and will be injured, more or less, in their thriftiness and productiveness, by other trees growing between. But, as the trees are many years reaching that size, and remain quite small for a time, reliable pomologists are inclined to favor this practice, where it is desired to make the most of the land for fruit-trees, provided a continued and enriching system of cultivation is adopted, by applying manure, turning under clover, and an occasional broadcast dressing of lime or ashes. J. J. Thomas says, touching this subject, that the roots of the peach, as we have found by decisive experiments, extend quite as far as the height of the tree; and therefore they soon form a network of roots over a circle whose diameter is twice as great as the height of the tree. Apples do nearly the same; and consequently, in a few years after setting out the orchard, the whole surface of the ground is covered with the roots of each, a few inches beneath the surface. The peach-trees will, therefore, soon interfere with those of the apple, and necessarily retard them. The apple-trees, after the lapse of a few years, will be lessened in growth by the peach-trees. But the latter probably will have paid for this damage by their crops of peaches; and generous manuring may nearly compensate for the loss. At all events, we have known the practice to succeed well on good soils; but, as the apple-trees advance, it would be best to keep the peach-trees well shortened back.

Dwarf pears will prove of no injury whatever to young apple-orchards; for the dwarfs must have, of necessity, good cultivation and manuring, which will ultimately benefit the apples. The roots of the dwarfs are short, and never extend very far; and consequently will not soon interfere with the apple roots. They will, however, suffer serious loss from the overshadowing influence of the apple-trees as the latter reach the age of sixteen or twenty years; and the dwarfs must not be expected to succeed after that. If in ground by themselves, the few pears that succeed best as dwarfs may be expected to flourish, with proper management, at least fifty years. We would not in any case recommend standard pears or plums among apples.

Planting an Orchard on New Land.—In many instances it is desirable to plant an orchard as soon as the forest is removed, before the land has been stumped and ploughed. If the soil does not require under-draining, it will be an excellent practice to put out the trees or plant the seed as soon as the brush and other rubbish can be disposed of. We have sometimes seen fruit-trees planted a few feet from a brush-heap or a pile of stumps, which have been killed by the heat when such rubbish was burned. Occasionally the place for a fruit-tree will be found where a stump is standing. Let such stumps be taken out at once, and thus provide an excellent hole for setting a tree.

There is often a wonderful advantage in planting an orchard on new land when the ground is full of roots. By the time the roots of fruit-trees have got a good start, the roots of the forest-trees will begin to decay. Then the roots of fruit-trees will spread rapidly in the places of the old roots, the decayed material of which will promote a much more rapid growth of the fruit-trees than if they had been planted on old ground. This fact will explain, in a good degree, why fruit-trees flourish more satisfactorily in a new country and on land just cleared, than on old ground of the same quality.

Cutting off the Tap-root. — Read about the *Habit of Trees*, in Glossary. The recommendation to cut off the *tap-root* of every tree when it is to be transplanted, once became so imperative, that some pomological writers even went so far as to recommend placing a broad flat stone beneath every tree, so that it could not reproduce another tap-root. Most writers at the present day, and most nurserymen, will say: "By all means prune off the tap-root close to the stem." But such a practice is a serious delusion. Nature indicates, as has been already stated, that most trees, shrubs, and plants will flourish most satisfactorily, endure longer, and be every way better when they send a strong tap-root deep into the ground. And this fact is more particularly true when the subsoil is so porous that a long tap-root will readily strike downward as far as the stem grows upward. Were it not a habit of the tree or plant to send down such a tap-root, it would be advisable to cut it off. But many nurserymen will persist in cutting this root off at all hazards; because, they say, ten roots will push out at the end of the stump where the tap-root was severed, which will be more serviceable to the growth of the tree. The office of the tap-root is to supply the growing plant or tree with moisture in hot and dry weather, when the surface-roots can not absorb one drop from the soil. Cut off the bruised end, and encourage a tap-root to grow on every tree and bush. (See p. 90, Fig. 46.)

Lateral Extension of the Roots of full-grown Apple-trees. —If we could lift a large apple-tree with all its roots unbroken from the earth, probably, to our great surprise, we should see roots on every side reaching farther away, with a thousand open mouths, than the entire extent of the longest branches. Tull found that common turnip-roots extended at least three or four feet from the plant; and there is no doubt that many weeds throw out their fibres quite as far. A. J. Downing, in his directions for the management and manuring of trees, indicated that the roots of

a tree were contained within a circle no larger than the
spread of the branches.   J. J. Thomas says that the roots
of common grass are found to run two feet, and sometimes
much more.   If, then, the tree-roots are but two and a half
feet long, making the circle five feet, they would meet and
come in contact with grass-roots two feet farther off, or
four and a half feet from the tree, requiring a circle of nine
feet at least to prevent the exhausting influence of the
grass.   When the tree is two or three years older, this circle would need to be doubled in size.   In other words, the
whole system of digging small spaces, and mulching small
circles about young trees, possesses very little utility when
compared with broadcast cultivation—the only mode that
should be adopted in orchards and gardens.   A tree, therefore, ten feet high, whose branches cover an area five feet
in diameter, reaches with its roots the circumference of a
circle at least twenty feet in diameter.   The *area* of the
twenty-feet circle is sixteen times greater than that of the
five-feet circle.   Thus Downing, in his directions, omitted
at least fifteen-sixteenths of the space covered by the roots.
There are many cultivators who do not keep the whole surface mellow, that loosen a circle of soil one-half as large as
Downing directed.   Nothing is more common than to see
trees ten or fifteen feet high with a pile of manure a foot
or two in diameter at the base of the trunk.   The tree can
not, as a matter of course, derive any more benefit from
such an enriching than a starving man could be fed by
stuffing crackers into his boots, with his hands tied behind
him.   If there is any one thing of great importance to fruit-
raisers, and which they are especially deficient in appreci-
ating, it is a knowledge of the wide extension of the roots
of trees.   A greater number of essential operations in their
culture is founded on this knowledge, than on every thing
else connected with vegetable physiology.   Many cultiva-

tors *know* that trees throw out their roots to long distances; but with this knowledge merely they seem entirely satisfied. They only *assent* to it, and let it pass. They seem to have no practical appreciation of it. Their experience in cultivation seems to have no connection with such knowledge; and we should naturally suppose, from their actions, that they thought trees ten feet high had roots only fifteen inches, instead of fifteen feet long. It may be laid down, as a general rule, that the roots extend in each direction from the foot of the tree as far as its whole height, and in many instances much farther. We see proofs of this fact where such trees as the locust and silver poplar throw up suckers at great distances from the trunk. The nurseryman who passes between his rows of salable trees is not often aware that the whole surface of the ground beneath his feet is covered with a net-work of roots, often extending the breadth of two or three rows. When he digs the trees by placing the spade a foot from the stem, he does not know that he cuts off and leaves nine-tenths of the fibres in the ground. The planter who sets them out supposes that for several years the roots only occupy a small circle, which he may spade and enrich, and thus afford them all the cultivation that is necessary. The roots of growing apple-trees require quite as much lateral space as the longest branches. It is too common an error to suppose that if the branches of trees have light and air enough, nothing further need be provided. But the facts already stated show that the tops may be far from meeting, and yet the roots may long since have become interlaced. Witness the extent of roots, p. 90, Fig. 46.

' All intelligent pomologists and tillers of the soil, as far back in ages past as we have any account, have appeared to understand the importance of preparing the soil for all kinds of trees in such a manner that some of the roots will

strike downward as far as the top extends upward. Hence
Virgil, who wrote nearly two thousand years ago, when alluding to this subject, says:

"High as his topmost boughs to heaven extend,
So low his roots to hell's dominion tend."

Removing large Trees.—It is desirable many times to remove trees which are twenty feet or more high, the bodies
of which are four to six inches in diameter. Winter is the
time to do this work. Some kinds of deciduous trees, like
the maple and elm, which are not so dependent on a tap-root
as many other kinds of trees, may be removed with more
assurance of success than many of the nut-bearing or cone-
bearing trees, which will not, in some instances, flourish at
all satisfactorily after the tap-root has been disturbed. As
maple-trees do not strike their roots so deep as many other
trees, large maples may be removed in cold weather with
less difficulty than almost any other kind of growing trees.

The correct way to take up a large tree is to remove the
leaves and loose mould on the surface around the tree, and
dig a channel about six or eight inches deep around the body,
about three or four feet from the centre of the tree. Then
fill this channel with leaves, and press them down close, to
exclude the frost from a small portion of the earth between
the ball that is to be removed with the tree and the remaining banks. This channel must be cut before the ground
freezes. Then, after the ground has frozen six or eight
inches in depth, take up the trees, set them on a sleigh, or
stone-boat, haul them where they are to be planted, and
drop each tree in a hole previously prepared. Many trees,
when removed in this manner, will continue to grow as luxuriantly as if they had never been taken up. A great many
new houses, having no shade-trees near them, can be provided with large trees at a small expense, if suitable trees
can be obtained within a distance of two miles.

By boring an inch hole through the body of the tree, to receive a large clevis-bolt, we have lifted trees alone, with a system of gin-poles and tackle-blocks, which would weigh —the ball of earth and the tree—more than one ton. The hole in the tree should be filled with grafting-wax, to prevent decay, after the tree has been planted in its desired locality. By having a windlass for winding up the slack rope of a set of tackles, one man can lift any tree that it would be safe to take up. Some care should be exercised that the ball is not planted deeper than the tree grew. As soon as a tree is dropped in its place every hole should be filled with earth; and it would be well to spread a few inches in depth of straw over the roots, to protect them from injury during very cold weather.

Fig. 49.

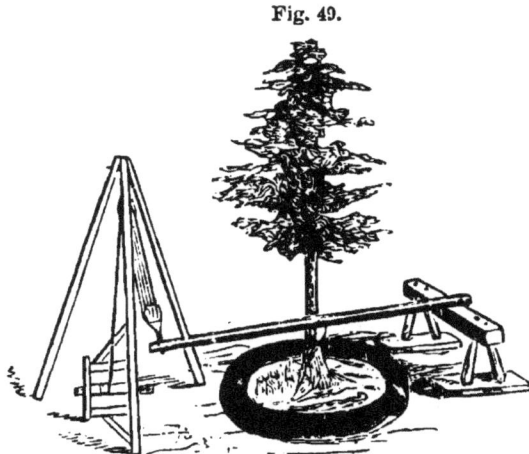

Apparatus for removing large trees.

The accompanying engraving (Fig. 49) represents the apparatus which we employed in removing large trees. It consists of a set of shears about sixteen feet long, with a windlass attached to the single shear, around which the slack rope of the tackles is wound when the machine is worked. A good stiff pole or stick of timber, twelve or fourteen feet long, is fastened to the tree with a large clevis and chain, by boring a hole through the tree for the bolt of the clevis, as near to the ground as can be. It will be difficult to hitch a chain around the body of a tree without wound-

5*

ing the bark. One end of the pole is supported by a strong bench three feet high, standing on planks to prevent its sinking into the ground. With the tackles hitched to the other end of the pole, the tree is lifted high enough to allow a sleigh, or stone-boat, to be backed under it, to receive the tree with the ball of earth.

When trees were to be removed, our practice was to dig a trench about them, about eight or ten inches deep, or more, according to the size of the tree and extent of the roots. This was done one year before the trees were removed, so that the roots in the ball of earth might send out a large number of minute fibres, which would not be there if the trees were removed as soon as the roots were severed by cutting the trench.

The next thing of first importance is, *to have the holes dug before a tree is brought to the ground.* Even if they are filled with snow, it can readily be removed when trees are ready to be placed in them. Holes should always be dug sufficiently large to receive the ball of earth without resting on the edges; and care should be exercised to have the trees no deeper in the soil than they naturally grew. Trees of considerable height should be stayed up with four guy-wires, to prevent the winds from blowing them over. These guy-wires should be fastened to the tree with small staples driven into the tree, and to stakes driven into the ground. When trees are small, we may dispense with the pole and bench, and hitch the tackle directly to the tree; and in *unloading* trees, we seldom used the pole and bench.

We once removed several evergreens from the forest, from twenty to thirty feet in height, and four to six inches in diameter; and those which were unloaded directly into holes already prepared are now alive; while all of those which were unloaded on the north side of the house, and the roots covered with moss and straw until the ground

POUND ROYAL.

Synonym.—Winter-pound Royal. The fruit is large, roundish-oblong-conical, with a surface slightly uneven; color, pale, yellowish-white, occasionally with a faint blush, and, when ripe, marked with a few large, ruddy, and dark specks; the flesh is very tender, breaking, fine-grained, with a mild, agreeable, aromatic, sub-acid taste; season, December to April. The tree has a broad spreading top, and requires the best of cultivation to insure good crops of fruit.

thawed—as we had seen recommended in some agricultural papers—died the same season. When the machine is
hitched to a tree, and we have lifted on it almost enough to
start it, take a lever or crowbar, and loosen the ball of earth
a little all around the trench. This precaution will sometimes prevent the ball of earth from breaking, and the breaking of some part of the machine also. With such a machine, we have often gone two miles from home, and re-

turned with a large tree and ball of earth six or seven feet in diameter and eight inches in thickness, loading and unloading *entirely alone*, in two hours and a half. We mention this fact to show the efficiency of the machine, and that it may be handled and worked advantageously by one man who knows how to handle such an apparatus.

Instances are of common occurrence where it is desirable to remove valuable fruit-trees in the summer or late in the spring, when we can not avail ourselves of a frozen ball of earth. When the writer was about seventeen years of age, a valuable pear-tree was to be removed, in order to make room for the wood-house. It was so old and large, that every one thought it absurd to attempt to transplant it with any expectation of its living. We had eaten too many delicious pears from that tree to see it cut down, which was the orders; but, in laudable obstinacy, we took it up in the middle of June and transplanted it, and it bore pears the same year, and is a valuable bearer even at the present writing.

We have seen a great many valuable fruit-trees cut down which might have been taken up and transplanted with very little injury to their productiveness. When a few tall shade-trees are wanted at once, such an apparatus will be found of great convenience in taking them up.

Removing large Trees without Tackles.—A farmer in Western New York recently came in possession of a large young orchard, the trees of which were standing in the quincunx order, and were about as thick again as they should stand. He removed every other row—over four hundred trees, from seven to ten years' standing. He wrote that the expense of removal was about forty cents each. He employed long levers and a broad stone-boat, and three men to load and unload the trees. According to his own account, the labor was performed in a very rough and unskillful manner; and yet only eight trees in every one hundred failed to grow the next season. If such an apparatus as Figure 50 had been employed, and if a circle had been cut around every tree the previous spring, the cost of transplanting would have been much less per tree; and not one tree would have died.

A Tree Machine on Wheels.—If a person were to have twenty or thirty trees to remove, the expense of making a frame, to be placed on the wheels of a common wagon, would cost but a few dollars. Two pieces of scantling, say three by six inches square, for sills, could be placed, edge up, on the bolsters of a wagon; then two windlasses should be fitted to the upper side of the sills, as shown by Fig. 50, around which chains may wind. The windlasses should be made of some hard and tough timber, about six inches in diameter, with two one and

Fig. 50.

Removing trees with a frozen ball of earth.

a half inch holes bored at right angles through each windlass. When a tree is to be lifted, remove the windlasses and one sill from the wagon, and place the vehicle in position with the reach or coupling of the wagon close to the body of the tree. Then lay the sill and windlasses in their places, and lift the tree clear from the ground, and drive carefully to the place for transplanting. An inch auger-hole bored through the body of a tree, to receive the bolt of the clevis, will not injure the tree in the least, if the hole is afterwards filled with grafting-wax. We have bored a hole through many trees five or more inches in diameter, when about to remove them, and we have never known one to be injured by the hole.

Transplanting as Trees grew. — It has been asserted, times without number, that trees should always be marked before they are taken up, so that they can be transplanted just as they grew originally, with respect to the points of the compass. Some philosophers, within a few years past, have stated as a fact that there is an unknown influence operating on fruit-trees, which induces the tops to incline towards the south east. We have been told, also, that people should always sleep with their heads due north. All these things *may* be of importance. But we do not repose any confidence in one of the suggestions; and it is extremely doubtful whether any one actually does know any thing positively with regard to the *facts* in the case. Many trees have been turned half way around when they were transplanted, and they produced bountiful crops. But no person has ever been able to affirm that *he knows* that some or all of those trees would have been more productive if they had been transplanted exactly as they originally stood.

CHAPTER V.

PRUNING AND TRAINING.

Cautious the pruner pinches from the second stalk
A tiny pimple that portends a future sprout.
None but his steel may touch the twigs doomed to the knife.
The limbs at measured distances, that air and sun
Admitted freely, may afford their genial aid,
To ventilate, and warm, and swell the fruit and buds.

COWPER'S *Task.*

The Philosophy of Pruning.—Pruning and training fruit-trees correctly is the most difficult of all the operations connected with the propagation of trees and the management of orchards, as no set of rules or recipes can be given for the pruning of fruit-trees which will enable a person unacquainted with the principles of vegetable growth to become a successful practitioner. A tree is not simply an individual organism or unit, like a man or a horse—it is a "Mutual Benefit Society," composed of a number of individuals, amounting sometimes to many millions, each one being capable, under favorable circumstances, of maintaining its own existence, not only when in connection with, but when separated from, the community in which it was produced; or it may easily be transferred to another society, and will there grow and reproduce its kind with undiminished vigor. Hence, for any one to tell on paper when is the proper time to prune a tree or not to prune it, under all circumstances, would be a task which has never as yet been done, and which we do not expect to perform in this place. But it is not difficult to state what effects usually follow pruning at a given period when different parts of a tree or plant are pruned. The cultivator should have a

perfect understanding of what he desires to accomplish by
pruning. Before he severs a single bud, he should understand the laws of vegetable physiology, so far as they affect
the flow of sap. There is no chance for correcting bad
mistakes in pruning. If one pinches off a bud that should
remain, it will be ruinous to the form of the tree or bush.
The practice, with a great many persons who have had the
management of orchards, has long been, and even now is,
to allow the trees to grow at random for several years, and
then walk into the tops with axe and saw, removing half
the branches. There was never a more ruinous practice to
any kind of trees; and there was never a more egregious
error promulgated, than to allow a bush or tree to grow at
pleasure for a few years, and then give it a thorough and
severe pruning. The ruinous consequences of such pruning are manifest, wherever large apple-trees are found, in
the decaying trunks at those points where large branches
were cut off; and because the wounds were so large, Nature could not heal them.

Pruning Trees of any Kind is a Science.—

> The novice at pruning, like urchins at school,
> With long observation must study each rule:
> Why prune in the summer? Why prune in the fall?
> Or why in the winter? Or why prune at all?—Edwards.

When a surgeon is about to amputate a man's limb, he
can assign a satisfactory reason for the operation. So
when a pomologist is about to prune off certain branches,
he should first ask, What do I propose to accomplish by
removing this part of the tree? Sometimes, when he thinks
that the removal of a branch will improve the general shape
of the tree, let him bend the branch a little out of its natural place, so as to show how the tree would look if this
limb were cut off; and if the result is not satisfactory,
don't cut it. There is a tendency to trim too much. Beware of this overdoing the matter. There was a time when

excessive pruning was very fashionable even among professional orchardists; but they have learned better. The current of opinion is setting so strongly the other way now, that some are advocating no pruning at all. Probably this will be found as much the other extreme; and in due time we may expect the true mode to be settled upon.

When we make a peculiar motion with an ox-whip, well-trained oxen understand the movement, if not a word is uttered. The motion means something. When a horse feels the whip, the blow of a good driver has a peculiar signification. A blow from an axe, a cut with the pruning-saw, a slash with the pruning-knife, and a pinch with the thumb-nail, all have a peculiar signification as affecting the growth of the tree and the development of branches. The growth and development of an apple-tree are so completely under the control of *law*, that a skillful pruner can make the branches assume almost any desired form. If he pinches or cuts wrong, the branches will grow wrong. If the pruning and pinching are performed in accordance with the law which controls the flow and circulation of the sap, every branch and twig will spread out, or stretch upward, just as the flow of the sap has been directed in a lateral direction, or vertically. Those persons who understand the philosophy of pruning, know that by pinching off the terminal bud, the upward, rampant growth will be checked, and the shoot will begin to enlarge in thickness and to increase in strength, and to send out lateral branches where nothing but buds existed. The true reason for this phenomenon is, the sap, which was before strongly attracted to the leading stem, is now distributed more equally among the other branches.

If we wish to make a tree or shrub "grow low," and extend its lateral branches in every direction as much as possible, we pinch off the tip ends of the leading, vertical stem, and thus induce the sap to flow more abundantly in the lateral branches. Young grafts, or inoculates, often shoot up two or three feet, if allowed to grow long, slender, and fragile; and such are liable to be destroyed by the first strong wind, unless they are supported by a stake. If it is our aim to make a tall tree by promoting the growth upward, we have only to cut off the lateral branches all round this leader, and thus induce as much of the sap to feed the leading sprout as is possible. But there is great danger of cutting off too many of the lateral branches. It would by no means be attended with the best results in promoting the growth of a tree upward, to trim it to a bare stem, leaving but a small bush near the top. If all the lateral branches are cut off close to the leading stem, and only a very small top left, if there be a strong healthy root, that very small top can not use up all the sap, and a *reaction* will follow; and sometimes such a reaction is attended with deleterious consequences to the tree. On this point thousands of young farmers, in their eagerness to promote the upward growth of their young fruit-trees, by trimming small trees to a single stem, with nothing but a few buds or bunches of leaves near the top, have found that their trees grew slowly when every thing—aside from this excessive pruning—favored a rampant growth. The truth is, if a tree has a healthy, strong root, the leading stem will grow upward much faster by clipping off the ends of the lateral branches—unless it has a very heavy top—than it would if we cut off most of them close to the stem. The leaves of a tree are its lungs; and there must be leaves on a tree in proportion to the sap that is sent upward by the roots. When there are not leaves enough to elaborate the sap, stagnation of the sap ensues in some trees and shrubs; while in others, numerous buds from the body of the tree will start, and long slender sprouts will, if allowed to grow, soon form a heavy top. By continuing to pinch off the extremities of the branches that are pushing upward, the lateral branches can thus be induced to spread out in a direction to make a low-topped tree. Before young tillers of the soil can be expected to understand *why, when,* and *how* to prune a fruit-tree correctly, they must read and think, and think and read, and make inquiry and observations on this subject for years.

Why we Prune.—We prune trees and vines to promote fruitfulness, to prevent the production of much small fruit, to produce fairer and larger fruit, and to make trees grow of a more desirable form and symmetry. At one time, we prune or pinch to induce a tree-top to spread out broader and to produce a lower top. At another time, we want the branches to push upward rather than laterally. Sometimes Nature sends out several rival shoots, all of which can not be fully developed; or, if they could, it would not be desirable to allow so many to grow. Hence a part must be cut away. When trees and vines are permitted to grow without pruning, it frequently occurs that four branches are formed where there should be only one. Hence, after a few years, one or more of these branches must fail and decay. To appearance, Nature made a mistake in starting too many branches. Hence she must divert all the sap into one or two, and dispose of the supernumerary ones as best she may. Nature does a vast amount of pruning, both in fruit-trees and in the forest. Branches are often allowed to decay and drop off the parent stem, because they were not needed. But such a mode of pruning is far from being scientific and artistic. And yet, it is the only way that Nature can perform this important operation. Hence we prune to aid Nature in performing her task as perfectly as possible. But pruning trees and vines is an operation requiring the skill of the pruner in connection with the efforts of Nature. She often starts a score of buds on a stem, only a small portion of which should be permitted to grow. Here the judgment and skill of the pruner must be brought into exercise just as the buds are beginning to develop. Nature has afforded an excellent opportunity to choose one or more of the most desirable buds for the future branches, and to destroy the others before the energies of the growing tree have been largely employed in producing large branches, which must be cut off as soon as they are formed. Some American pomologists of the present day advocate the utter abandonment of pruning fruit-trees; and they have pointed to productive trees of beautiful form and symmetry as evidence that Nature does not need the assistance of art in the production of fruit-orchards. But it is never safe to leave any tree entirely to the operations of Nature, as some trees will grow about as nearly correct in every particular as can be desired, while others will assume the density of a thicket, and others still will send out long, slender branches and twigs, having only a few fruit-buds at the extremities, reminding us of a far-reaching and covetous person for the acquisition of a broad extent of country, which he has no ability to occupy or cultivate. Nature will sometimes do all necessary pruning; but she will be a long period doing it. Hence we pinch and prune, to aid Nature both in facilitating her operations, and in performing her allotted tasks in a skillful and artistic manner. As has already been stated, many trees will grow and develop almost every branch and bud in a symmetrical and artistic manner. So, many children seem to grow up to manhood and womanhood complete ladies and gentlemen, without any apparent training. Still, there is a law in fruit-culture which requires the exercise of the cultivator's judgment in pinching and pruning; and he who undertakes to evade or ignore it in the neglect of his fruit-trees will find his well-deserved punishment in miserable fruits.

Beginning to Prune.—We have, for example, a young grafted tree (Fig. 46), the top of which it is desirable to train, from year to year, until a neat and symmetrical head is formed, similar to Fig. 51. During the first two to four

Fig. 51.

An apple-tree properly trained and pruned.

seasons, according to the rapidity of the growth of the young tree, no other pruning-tool than the thumb-nail will be required, providing the pruning is done at the proper time. All the low branches, spurs, and leaves should be allowed to grow, as a young tree can not flourish without leaves. But, after the branches have grown about one foot long on each side, let the ends be pinched off, say one-fourth of an inch, sufficient to check their lateral extension, to cause the stem of the tree to grow more stocky, and to induce the sap to push the central stem upward with more rapidity. But, while it is the aim to promote the upward growth of the central stem, the pruner must exercise judgment over the upward growth, that it does not shoot up too tall and slender. If the central stem does not develop in size and lateral proportions fully equal to the upward growth, let the top be pinched a trifle, which will induce the head to thicken-up. In case the topmost bud or buds start upward before the body of the stem and side branches have become sufficiently stocky, pinch the bud alluded to again, and keep the upward growth back, and thus promote the thickening of the main stem. Many young pruners, in their eagerness to get tall trees, have rubbed off all the buds on a slender stem, except a small number near the top; and the consequence was, that the growth was very slender and feeble, so much so that there was nothing but a central stem, and that was so slender that it could not stand erect without a stake. In order to have the central stem grow with the greatest rapidity, we must reserve stems and leaves around it, clear to the ground. But the pruner must avoid the error of promoting a long and slender growth to the neglect of a stocky development. If the central stem does not enlarge in size, so as to be of a fair proportion to the height, keep the topmost bud pinched off, and thus stop the upward growth entirely until every part is thickened-up in a satisfactory manner. Then the central stem may be allowed to push up another foot. The great aim is to get every part started correctly.

Low or High Heads.—If it is the purpose of the pruner to form low heads, so low that one can step from the ground on the first branches, let him select four branches, or buds, say two feet from the ground, to be trained as the first system of lateral branches. About twenty inches above these choose four other buds for a second system of branches;

and so continue to do as the central stem pushes upward, rubbing or cutting off all buds and twigs on the central stem between every two systems of branches. The central stem will push upward two feet or more annually, if the soil is fertile and properly cultivated. But it will be more satisfactory, every summer after the central stem has pushed upward about twenty inches, to pinch the topmost bud, and stop its upward growth for the season. Then, the next year, it will send out a system of lateral branches at the point where the growth of the previous season ended. In case it is desirable to form high heads, so high that a horse can travel beneath the lower branches, let the first system of limbs be commenced about five feet from the ground, after which proceed according to previous directions. All through the growing season, the pruner must watch his young trees every week, to see if the tops are growing uni formly on every side. One branch may be pushing out too rapidly, and robbing its fellow of a proper proportion of sap and space. Let the terminal buds of such aspirants be pinched a little, and continue to pinch them until others are even with them. When too large a number of twigs push out on the sides of lateral branches, cut off those that will not be needed. If a sub-branch begins to push away more rapidly than those around it, pinch the end, to keep its growth within desired bounds. By attending to pinching at the proper time, a pruner will be able to make his tree-- tops assume almost any desired form.

How to Prune.—Before a person proceeds to pinch off a bud or cut off a branch, he should be able to state to an- other person, what kind of a tree, as to form, he desires— what he proposes to effect by pinching off a bud or sever- ing a branch, and *how* his tampering with buds and branch- es will accomplish the end in view. The first important point is to obtain a definite and correct idea of what you

TETOFSKY.

The fruit is of a medium size, oblate-conic, frequently round, or nearly so, and smooth, with a yellow ground beautifully striped with red, covered with a light-colored bloom, beneath which is a shining skin. The flesh is white and juicy, sprightly acid, fragrant and agreeable; season usually, at the North, August and September. The tree is an upright and spreading grower, a good bearer, and produces fair crops of fruit when only a few years old. It is usually a very hardy variety, and the leaves are large and heavy.

wish to do before you touch the tree. A general vague conviction that fruit-trees need pruning, or thinning out to keep an open head, by removing weak and conflicting branches, constitutes the whole stock of information with which most persons commence the yearly attack upon the orchard. There is no careful study of the habit and peculiarities of each species of tree; no thought of what each individual tree has done in the past, or is expected to do in the future, whether it is prematurely forming fruit buds, or running to wood too luxuriantly; no special care for a

Fig. 52.

Fig. 54.

Pruning - saws.—When pruning trees, the branches of which are too large to be cut off with a strong pruning-knife, a pruning-saw or two (Fig. 52) will be desirable. A and B represent the most approved style of such saws, copied, by permission, from R. H. Allen's Catalogue of Agricultural Implements, one of which has coarse teeth, for cutting off large branches, and the other fine teeth, for small branches.

Fig. 53.

Pruning-shears.

Pruning - shears, like the above (Fig. 53), will frequently be found convenient when clipping the extremities of small branches.

A Combined Pruning - chisel and Saw.—Figure 54 represents a convenient instrument for pruning old trees, while the operator stands on the ground. The end of a stiff pole enters the socket; and small branches may be severed with the chisel by means of a thrust, or they can be sawed off at the desired place. Such a combined instrument will be found efficient for removing sprouts and dead branches from neglected trees. The saw should have fine teeth, or it will be difficult to make it work. In some instances, the saw will cut more satisfactorily by changing ends with it.

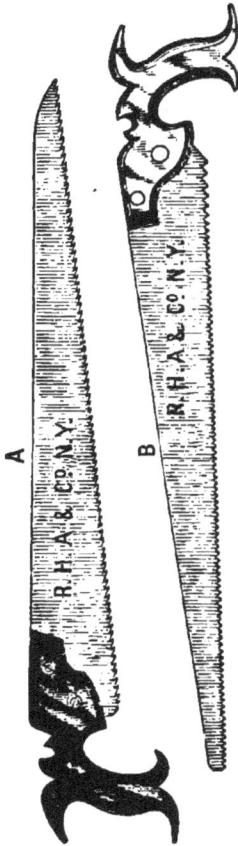

Pruning-saws.

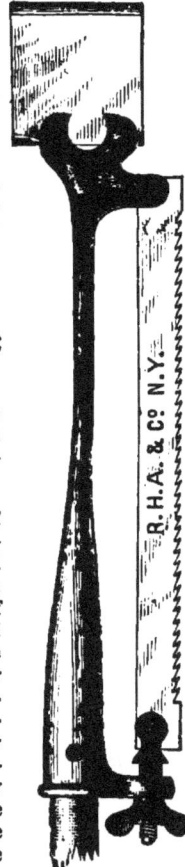

A pruning-saw and chisel.

weak but important shoot which is receiving too little nourishment, because a gourmand above it is monopolizing all the sap and sunlight; no calculation for future years, that the foundation now laid shall be the basis of a sufficient number of branches, filling advantageously every part of the tree, while none shall crowd or interfere with its neighbors.

In most instances, the tree is said to need pruning, and

the attack is made. The saw and axe are brought, and in a single hour one-third of the top is cut out. A tree should never, in this sense, *need pruning.* The difficulty should be avoided, rather than remedied; so that, instead of felling great branches, the finger and thumb, or at most the pruning-knife, will be sufficient to *direct* the growth of stalwart limbs. The direction of a skillful pruner always is to fix in your mind the general form of a perfect tree of the variety you are about to operate upon, and to this ideal, as nearly as possible, train your subject; not, of course, arbitrarily, or in one year, but, by patiently studying the peculiarities of your tree, bring it gradually to the desired form. In respect to the shape, fruit-trees may be classified into *globular,* or round-headed trees, like the apple; *semiglobular,* or goblet-shaped, as the peach; and the conical, like the pear and the cherry. It is important that, while we divert Nature from her wonted course to fulfill our especial ends, we do no violence to her *principles.* Faults there are to be corrected, deficiencies to be supplied, but always obediently to the guidance of nature. There is a typical form, then, for each variety of tree, which should be regarded from the commencement. A pruner must set up before the mind a *beau idéal* of the form of tree or bush desired. Then, all through the growing season, the buds should be watched closely. If a bud appears where a branch is not desired, pinch it off; and leave buds on the main stem wherever a branch is desired. Grafts from cions that were set last season should be examined frequently, to see if the main stems and lateral branches are all growing uniformly. The pruner must understand the *manner of growth* common to the variety he is about to prune. All apple-trees do not grow alike. Hence, for a pruner to attempt to compel a tree having a natural tendency to a certain form to assume another, would be attended with much un-

Fig. 55.

A strong pruning-knife.

availing trouble, and probably with positive injury to the productiveness of the tree. The great difficulty is to make the lower branches grow thrifty, and in due proportion to the upper ones. The whole secret lies in *the management of the buds.* Every shoot and branch commences life as a bud; and it is in infancy that their proper number and position must be determined. Leave no more buds upon a shoot whose growth you wish to increase than can be maintained in perfect vigor. This will generally be about one-third of the number of buds produced; so that of those shoots designed to receive the largest development, two-thirds of the last year's growth must be cut off. These should be shortened-in before they start in the spring. If still the upper branches grow too strong, summer *pinching* will furnish the requisite discipline for them. We have always found it a safe and correct rule, when pruning a fruit-tree of any kind, to imagine that a full-grown apple-tree was standing before us, with no supernumerary branches, and no limbs crowding each other, or riding one on the other, like the cut of a tree, Fig. 56, which represents an apple-tree that has been train-

Fig. 56.

An apple-tree trained with a high top.

ed from its first year's growth until it had developed into a perfect tree that needs no pruning. The top, as may be seen, is well filled up with bearing branches; the form is symmetrical, the head high, and the pruning has been such that a person can climb about in the top, when gathering the fruit, with little difficulty. This is only *one* model. Although every apple-tree can not be expected to have a top exactly like this, unless it is of the same variety, still those that grow differently may be trained so that the form will be quite as symmetrical as this. One great aim of the pruner must be to have the middle of the top filled up with bearing branches. Then he must avoid pruning in such a manner as to have wide-spreading tops, with only a large bush on the end of a long branch.

When to Prune.—One respected pomologist, in reply to such an interrogation, will respond, "Whenever your knife is sharp"—implying that doing the work well is more important than the selection of any particular period. Another authority declares: "The season for pruning is usually midwinter, or at *midsummer*. It is, however, the practice to perform what is called the winter pruning *early in the autumn*." An author who has penned about twenty pages on pruning, is so diffuse and indefinite, that all a beginner can gather is, "Prune in winter for wood, and in summer for fruit." A. J. Downing, whose writings are repeatedly quoted as reliable authority on fruit, says: "We should especially avoid pruning at that period in spring when the buds are swelling and the sap is in full flow, as the loss of sap by bleeding is very injurious to most trees, and in some brings on a serious and incurable canker in the limbs. Our experience has led us to believe that, practically, a fortnight before midsummer is by far the best season, on the whole, for pruning in the Northern and Middle States. Wounds made at this season heal over freely and rapidly."

Another reputed authority says: "The best time for a general pruning is at the close of the first growth of summer, 15th of June to 15th of July." Another pomologist says : "*June* is the time to prune fruit-trees. Limbs taken off at this season will begin immediately to send out a ring of new wood just where it is needed, and will thereby protect itself in the soonest possible period from external harm." Still another writer—whether he ever pruned a tree or not, does not appear—says : "From the middle of June to the first of September is claimed to be the proper time in which to perform this important operation. Most persons have observed that trees show in August and the early part of September what is called a *new growth*. On this growth the color of the foliage is a lighter green, and has every way the appearance of being more recent than that of the rest of the tree. And so it is. By the time that mid-summer comes, most of the sap that flowed up in the spring has gone to the branches, and aided in expanding buds and blossoms, and in sending out new leaves and extending the twigs. When the tree has done this, the superabundant sap returns down the tree through the bark, and increases its diameter. The tree has now a season of rest. The sap-vessels are comparatively empty, so that if its branches are cut the wound will rarely bleed. The returning sap, we suppose, soon forms a green, healthy ring about the cut in the bark, and the remainder of the cut dries and shrinks before the sap is again in motion. This season of rest, then, of three or more weeks, is the best time to prune. It has its inconveniences, we are aware, but they are of less consequence than the injury of the tree. No harm comes to the tree, we believe, if pruned in the autumn, soon after the leaves have fallen. The tree is then also in a comparative state of rest, and may be cut judiciously without injuring it." Still another assumed authority proclaims : "Always prune in winter, when there is no foliage on the trees." Another author, who knows how to prune trees far better than how to instruct an inquirer after truth, says: " The correct period of pruning is when the tree needs it." This is philosophically and practically correct. And yet the phrase requires an intelligent qualification which we will endeavor to record, so that a beginner may not be in doubt as to the proper period of pruning.

When a tree has been allowed to grow at random for several years, so that the top is like a thicket, it should *not* be pruned during the growing season; yet it may be pruned at any time after the leaves have fallen, until the buds expand in the spring. No doubt every practical pomologist will concur in this declaration, as it is always far better for a tree to remove large branches at that season of the year when the sap is in the least motion. But when young trees are to be pruned and trained, more or less pinching of buds, and removal of twigs with a knife, must be done during the entire growing season. Hence, to affirm that the best time to prune is when one has a sharp knife, is about as correct as the direction to train children when one is in possession of a rawhide or rattan. As young children require daily training from the cradle to manhood—not when they are asleep—so young trees require care and pinching of buds, if necessary, from the first year of their growth until their branches are bending beneath a load of fine fruit. When the young tree begins to grow in the spring of the year, if every bud and branch is not growing correctly, *then* is the time to pinch or prune, as the case may be. We do not let our dear children, when they first begin to walk, tottle off the door-step, run into the fire, or grow up in vice, disobedience, and immorality, and then give them a cruel chastisement for their conduct. But we begin to lead, direct, and mould *in the bud.* So we must begin in early spring-time to pinch off certain buds that would grow into branches which would have to be cut off. Here lies the great secret of scientific pruning and training—pinching off a bud that would make a branch which must be cut off. A pruner who is master of his employment, can glance at a young tree, see where there are buds, and in three minutes prune it with his thumb-nail, so that the tree *may* need no more care for the entire season. Again, another young tree, like a perverse youth, may require pinching a little every month during the growing season. These suggestions will furnish a beginner with some correct notions as to the *proper time* for pruning.

Trees which have been properly managed during their whole growth will never need the cutting away of large limbs, unless they have been injured by teams or broken by snow or wind. Pinching buds and cutting off branches will always depend so much on circumstances, that a beginner must first make himself familiar with all the operations of pruning before he will be competent to train fruit-trees of any kind. When he has learned *how* to prune correctly, he will have a more perfect understanding as to the most proper period during the entire year. He must bear in mind that we often prune to diminish nutritive vigor, and prune to increase it; to diminish the generative or fruit-producing tendency, and to increase it; to encourage the feeble, and reduce the over-luxuriant. By a variation in the same process, under favorable circumstances, we can paralyze the leaf-bud, producing thereby a blossom-bud; and we can stimulate the blossom-bud until a leaf-bud will be developed.

Where to cut off Branches.—Nature has indicated, even in the smallest twig, the proper place to cut off a branch. By examining a branch close to the body of a tree, it will be seen that there are creases, beads, or rings running around the branch. The place to sever a branch, there-fore, is close to the ring around the stub at A. Then the wound will heal much sooner than if the ring were re-moved, making the cut smooth with the body of the tree. Many tree-murderers often cut off limbs several inches from the main stem. This is the true and effectual way to make trees decay at the heart. This illustration (Fig. 57)

Fig. 57.

will elucidate that point with suf-ficient clearness to enable a pruner to see at a glance what most prun-ers have never thought of. If a tree has been properly pruned, there will be no occasion for cut-ting off large branches, except in rare instances. In case branches should be injured in any way, so that it would appear necessary to cut off the limb close to the body of the tree, the cultivator should know exactly where the right place is. A limb as large as a man's arm or leg may be sawed off a large tree, and the wound

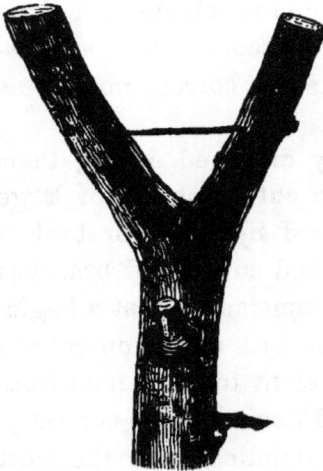

How to bolt a forked tree: where to cut off branches.

heal over readily, provided a clear and smooth cut be made, as indicated, and the surface be covered with a heavy coat of grafting-wax, and the wax covered with a piece of strong paper, as previously directed.

Some pruners have recommended, when taking off large limbs, not to make the cut parallel with the stem, but start the knife on the lower side of the branch, a little up from the base of it, and bring it out on the upper side exactly where the branch parts from the stem, so that the cut, when made, will form, measured down the trunk, an angle with it of 65 or 70 degrees. If a branch is cut off smooth, parallel with and close to the trunk, the wound, it is contended, will be one-third larger, with no advantage, and requiring a greater effort of the tree to recover. But by starting the cut a little way out from the stem, and cutting upward exactly to it, the wound is not only smaller, but sap will determine more strongly to the base of the cut, which will soon cause the wound to be covered. Such wounds should be pared smooth, in order that they may quickly heal over. Others will persist in cutting off large limbs from one to two inches from the rings around the base of the branch, as shown at A (Fig. 57). Such long stubs will almost always decay; and if they are large, the entire stub will rot clear to the heart of the tree.

How to Cut off large Branches.—Limbs of fruit-trees are sometimes broken down by snow or by a furious wind, so that it becomes necessary to cut them off close to the body of the tree. When the amputation of a large branch is important, let it be sawed first, at least one-third off, on the under side, say six inches from the central stem. Then run the saw down on the upper side, to meet the under kerf. After the limb has dropped to the ground, saw off the stub with a smooth cut at the rings, as directed p. 128, and cover the wound with wax at once. This precaution is necessary to prevent the branch from splitting down, like Fig. 58, before it is sawed off, and thus making a large wound on the side of the tree. Such large branches are frequently cut off with an axe, which will leave the wound very jagged;

Fig. 58.

Branch splitting down when sawed wrongly.

but if a sharp pruning-saw be employed, the surface of the wound will be smooth, and will quickly heal over.

6*

Fig. 59.

An apple-tree pruned wrongly.

Injurious Pruning.—Fig. 59 represents a style of pruning which has been, and still is, in vogue from Maine to California. The system may properly be denominated *tree - murdering*. Let this be compared with the well - pruned tree, p. 117, and it will be perceived at once that the system illustrated by Fig. 59 is exceedingly defective. And yet, this is a fair type of the form which the usual treatment of fruit-trees produces. It may be perceived that the whole growth of young wood and leaves is in the upper part of the tree-top. This occurs in obedience to a law of vegetable growth, which gives greater development to the terminal buds, and to those shoots which are nearest to the extremities of the branches. This tendency is very much increased by the pruning which has been practised by the cultivator, who evidently had a very indefinite idea of the objects to be obtained by this operation. Yet, having heard from his infancy that fruit-trees should be pruned, with such generalities for his guide in the way of instructions as "thin out the top," "take out weak or decaying branches," "keep the head of the tree open," etc., he has applied axe and saw to the limbs most conveniently reached, especially as he finds these to be the weaker branches. In

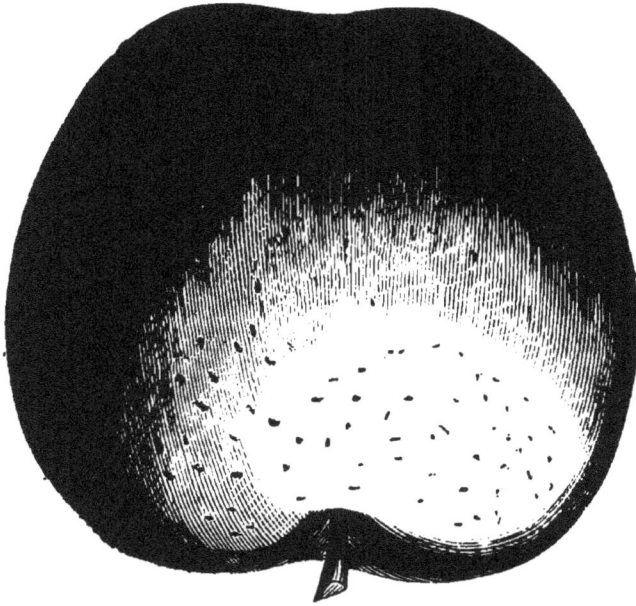

GRIMES'S GOLDEN PIPPIN.

Synonym.—Grimes's Golden. This excellent apple originated in Virginia; and in its native locality the tree is prolific and hardy. The growth is upright-spreading; the fruit is of medium size, roundish-oblate, and slightly conical; skin, somewhat uneven; the color is a rich golden yellow, sprinkled with small gray and light-colored dots; flesh, yellow, compact, crisp, tender, juicy, sprightly, sub-acid, with a peculiar aroma. In quality, it ranks good to best; season, December to April. A fine pie-apple.

many orchards we may see long, bare arms, utterly destitute of any twigs, fruit-spurs, or foliage, for eight to twelve feet from the body of the tree. The author has been in more than one apple-tree top, some of the large limbs of which were perfectly bare for fourteen feet in length! On the end of these long arms there were branches and twigs about equal to a small tree five or six years old. Is any person possessed of sufficient wisdom to tell of the advantage resulting from such long and bare arms? We have never met with such a savant. Nature always makes a desperate effort to correct the errors of a tree-murderer, by sending out numerous sprouts to shade the bare arms of the tree, as represented by the cut on p. 130. This fact shows that all the vacant space within the head of the tree should be occupied by bearing branches; and the upper side of horizontal arms should be covered with fruit-spurs and leaves, or small branches, to shade the surface of the large limbs. Many a valuable tree has been seriously damaged by the scalding heat of the sun pouring down on a large limb or on the side of a leaning tree.

Forked Trees to be avoided.—Many fruit-trees are more inclined to produce a forked than a single stem. A forked top should never be allowed to form, as such tree-tops are exceedingly liable to split apart at the fork, to the great injury of the tree, and frequently to its entire destruction. It is very easy to prevent the formation of a fork. When two small shoots become rivals, let one of them be kept back by pinching, and be trained as a lateral branch, or let it be cut off smoothly close to the other stem. A fork should not be allowed to form on any of the branches, as one or the other will most assuredly split down, either beneath the heavy burden of fruit or when some person steps on it, while gathering the fruit. There is no gain in any respect by allowing a tree to grow into a forked top. The tree will not be so symmetrical, neither will it yield any more fruit. Hence, if a fork has been allowed to form, so that the branches are as large as a man's arm, it will be far better to saw off one fork than to allow the top to increase in size. But see that the wound is covered with a heavy coat of grafting-wax and paper. In case a forked tree-top has attained a large size, let a half-inch iron bolt be put through the body six or eight inches above the junction of the forks, and let the nut be screwed up tightly. Let a large flat head be made on the bolt, and let a wide washer be placed beneath the nut. Should there be two lateral

branches which can be united by ap-
proach-grafting, the two forks may be
kept from splitting down by this arrangement. When pruning young trees,
let the pruner keep this one thought
bright, to avoid forked tops and forked
branches. (See Fig. 57, p. 128.)

**Doty's Combined Pruner and Fruit-
plucker.**—The illustration herewith giv-
en represents a useful and efficient tool,
which has been perfected by Wm. M.
Doty, of New York city, for pruning
berry-bushes; and which is an admirable edge-tool for pruning any kind of
fruit-trees. At the left side of Fig. 60,
the pruner and plucker is represented
as ready for use. At *a* the little sack
is represented, which is employed to receive an apple when the chisel severs
the stem. At *b* the knob on the end of
the piston is represented. The piston,
b, plays back and forth through the metallic sheath, *c*. At *e* an enlarged view
of the metallic hook is given, which appears in the upper end of the left-hand
figure. At *f*, a thin wide chisel is represented, which moves back and forth on
the side of the hook *e*, when the tool is

Fig. 60.

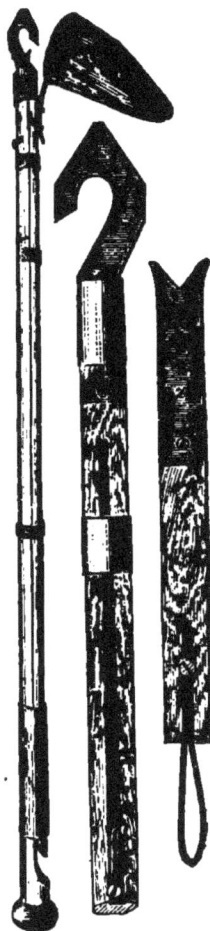

Doty's combined pruner
and fruit-plucker.

in use. At *g* a spring is shown, which draws the chisel back
after it has been thrust forward to cut off the branch that
may be received in the hook, *e*. A shoulder is formed on
the piston, near the iron knob *b*, by not dividing the pole
quite its entire length, which, striking against the end of
the hook-holding portion, arrests any surplus force of stroke

without injury to the machine. In case a branch is not entirely severed with one stroke, the chisel, being detached from the piston, remains at the point to which it was driven at each stroke, until through, when *the spring* withdraws it. By employing a long handle, such a device will be found convenient for cutting off twigs, shortening-in branches, thinning out tree-tops, clipping off cions, cutting out caterpillars' nests, pruning all kinds of berry-bushes and grapevines, and gathering choice specimens of fruit from high branches.

Training Fan-shaped Heads.—One of the most beautiful forms for training the tops of apple-trees, is to train the branches in the form of a fan, like Fig. 61, which represents one of a number of apple-trees trained in this manner by an amateur pomologist living on Long Island, whose trees, at the present writing, are about twelve feet high, and fourteen feet broad.

Fig. 61.

A fan-shaped apple-tree.

The gentleman stated to us that he entertained the conviction that those trees are not so productive as they would have been, growing in the same places, if the heads had been allowed to assume their natural form. But there were no satisfactory grounds for holding such a belief, as those fan-shaped trees were quite as productive as others near them. We recommend this form of training only for gardens, and other places where space is limited. Still, a person can produce more than twice the amount of fruit on an area of two rods square—four square rods—by having that amount of ground occupied by three fan-shaped trees, than one tree would yield if allowed to grow in its natural form.

The great advantage of training an apple-tree in the form of a fan is, all the branches, from the central stem to the extremities of the longest arms, can be filled with fruit, most of which will appear on the outside of the branches, where it will receive the full benefit of the sunlight and air. When of this form, the annual cutting back of straggling twigs or rampant branches can be performed more advantageously than if the head were round. Besides this, heads of such a form are more convenient than round heads, when hunting depredators, when thinning out the fruit, and when gathering it; as fruit-ladders can be placed close by either side, within reach of every apple on the tree. There is no doubt that it would pay—where ground is scarce—to raise

three fan-shaped trees on the same ground that is appropriated to one tree with a globular or conical head. In addition to the foregoing advantages, the trainer has all the opportunity that can be desired to renew the bearing twigs, by cutting them back.

Fig. 62 represents a young tree which is being trained with a fan-shaped head. The dotted lines indicate where the ends of the branches should be cut back, when they grow too long and slender.

Fig. 62.

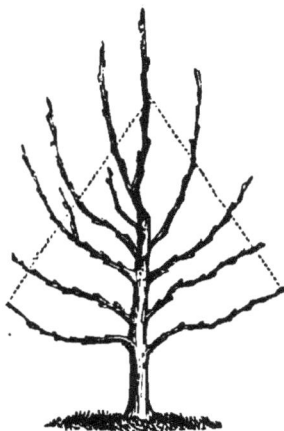

Training a fan-shaped tree.

How to Train a Fan-shaped Head.—The true way is to begin when the young tree commences the second year's growth, by leaving one bud on the south side of the main stem, say twenty inches above the surface of the ground for a lateral branch; then, nearly opposite of this branch, leave another bud, on the north side; and pinch off all others. Then, about twenty inches apart, on both the south side and the north side, leave buds to push out for branches, and encourage them to grow in the desired direction. If any buds or twigs appear on the east or west sides of the upright central stem, let them be removed. It is desirable that the branches extend nearly north and south, that the fruit on both sides of the tree may have the benefit of the sun more uniformly than if the branches were trained east and west. After the central stem has attained a height of six feet, it should not be allowed to grow upward more than one foot annually. After it has grown about a foot, pinch the terminal bud, so as to check the upward growth, and make the stem grow more stocky and strong. If any one of the lateral branches should push out faster than those above or below it, pinch the extremity, and keep the growth back, so that the branches above and below it may have an equal chance to spread. The branches will require attention every week, for the purpose of maintaining a uniform growth. After the lateral branches have begun to spread out, buds will appear on them, a portion of which must be pinched off. About six inches from the central stem, encourage a bud to send up a fruit-spur, and so train upright shoots six inches apart on the upper side of every lateral branch. Great care must be exercised lest some of the branches shoot outward or upward too rapidly. The entire space between all the branches must be filled up with fruit-bearing twigs. There will be a sufficient number of buds to form an upright shoot, as often as is desirable, if the lateral growth is not permitted to shoot away too slender. And the same is true of the upward growth of the central stem: if it is allowed to push upward ad libitum, in many instances buds will be wanting. In some instances, a branch may require a stay to hold it for a few weeks in the desired position. For this purpose pieces of lath may be secured to other branches first; then the perverse branch can be tied to the lath. It should be remembered that all the training alluded to should be performed during the growing season. By watching the first development of the buds, it will be an easy task to train the young and tender twig in the desired direction. But let a twig grow at random for a few weeks, and the trainer will need the aid of numerous stays to hold the branches and sub-branches in the desired position. The foregoing suggestions are particularly applicable to young trees that have sprung from the seed, where they are to grow, or to any young tree which has not developed any branches.

Training Nursery-trees. — There are two ways of training nursery-trees with a fan-shaped head, one of which is to remove the entire top when the tree is transplanted, and train it as directed in the preceding paragraph, as the new shoots push out from the stub of the tree; and another way is, to tie, or secure by pieces of leather, horizontal sticks for stays, extending across the top already formed. Then let the branches be spread out and secured by loops of leather put around the branches, and the ends fastened with nails to the stays. We once spread out the branches of an apple-tree twelve feet high, forming a fan-shaped head, fastening them to horizontal stays, and completed a satisfactory job. Some of the branches on the west side had to be removed entirely, and some of the vacant spaces on the upper side of the remaining limbs had to be filled by inserting cions by side-grafting. (See Fig. 5, p. 29.)

Fig. 63.

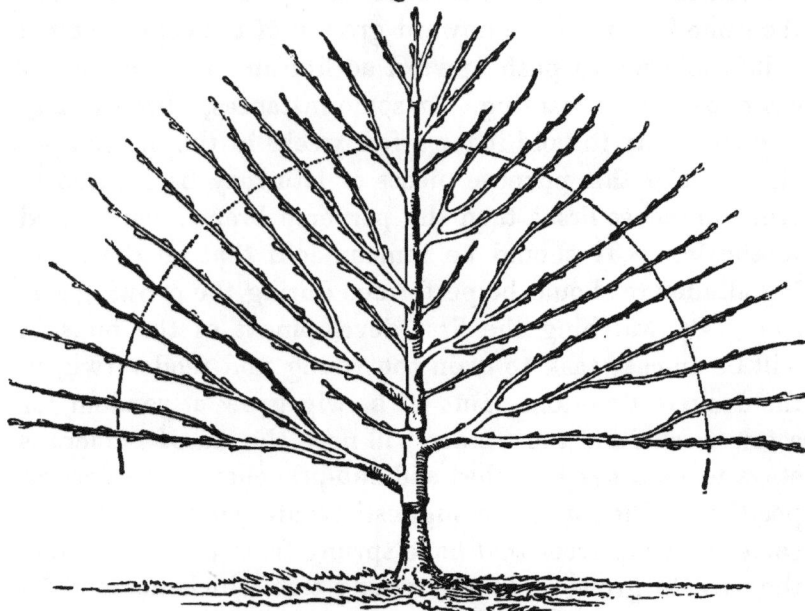

A fan-shaped tree.

Renewing Apple-trees.—The fan-shaped system, as was hinted at on a preceding page, enables the pomologist to renew the bearing branches with more facility than it can be done when the head of the tree is of any other form. When the bearing shoots have become too old to yield full crops, let a portion of them be cut away entirely, or in part, after which new shoots and new fruit-spurs will appear, which may be trained at pleasure, according to the directions of the preceding pages. Fig. 63 indicates, by the circular dotted line, where the branches may be cut back, when it is desirable to renew the bearing twigs. This is easily accomplished by summer pruning, done generally in July or the fore part of August, and even later, when the season of growth is extended. It is not necessary to prune severely to produce the desired effect; it will be sufficient to clip off

CANADA REINETTE.

Synonyms, numerous.—White Pippin; (erroneously) Yellow Newton Pippin. *Fruit.*—Size, large to extra large; form, varying, generally roundish flattened, slightly oblique, angular, much ribbed, especially towards the crown or calyx; color, light greenish yellow, with frequently a faint blush of red on the sun-exposed side: many small dark green specks, surrounded with light green suffused beneath the skin, slightly russeted; flesh, yellowish white, juicy, crisp, tender, sharp, sub-acid, sprightly, aromatic; core, small, compact; seeds very dark brown, almost black; season, December to May. *Tree.*—A very strong, vigorous, upright grower while young, forming a large, spreading, open-orchard tree, quite hardy, produc-tive, and profitable. An old French variety, described nearly two centuries ago.

the ends of the prominent shoots from time to time, as they may be pushing forth to excess. Pinching should be done late enough in the season to prevent after-growth, or should be kept up until such growth is impossible.

When it is desirable to renew apple-trees having globular or conical heads, the cutting back and thinning out must be performed by an experienced pruner, who undestands exactly what he desires to accomplish. In cutting back for the purpose of renewing the growth and improving the productiveness of a tree, we should endeavor not only to have the top well balanced, but open also, so that the air and light can have free access to all parts of the tree. A free circulation of these is necessary to vigor and vitality in the interior of the top; the want of it is what causes so many dead twigs around the body and large limbs of the tree. The fruit which grows in the centre of the top is often shut out from the light and air, and is consequently wanting in flavor and maturity. Some, on this account, have recommended an effort, by pruning, to throw the fruit wholly on the outside of the tree. We have seen what were undoubtedly many good attempts at this—trees where the interior was all cut out, and each limb had a likeness to the caudal appendage of a mule, the brush all at the extreme end. We think the better way is to thin out the twigs on the outside, and admit the light and air to the more sheltered portions. One advantage derived from this would be, that in seasons when late frosts or severe storms destroy the fruit on the outside, a portion of that within would probably escape injury. Whatever may be the style of pruning, there should never be any necessity for removing large limbs. Every branch not wanted should be taken off while small. It is impossible to grow apples fair and well-flavored on a tree crowded with superfluous branches. Its sluggish circulation and dense foliage, which excludes the sun, can furnish nothing but gnarly and insipid fruit. As the apple-tree always branches low, this inclination should be respected. Trees with low heads stand firmer, their trunks are less exposed, and their fruit is much easier gath-

ered. If the limbs come close to the ground, they operate
as a mulch, keeping the soil loose and moist. Trimming
should always be carefully done, the branch being smoothly
cut, without leaving a stub to die and produce decay, and
prevent the healing over. More than this, whenever a
branch over one-fourth of an inch in diameter is pruned
off, the wound made should be covered with wax the same
hour, or day at least, as previously suggested.

Root-pruning.—Unfortunately for the writer, he knows
only enough about law to keep out of cases of litigation;
and only enough about root-pruning to advise pomologists
to let Nature take care of the growth of the roots. Few
persons, probably, have investigated the subject of root-
pruning more thoroughly than we, having from boyhood
been on the watch for a single instance or experiment in
root-pruning which afforded even a *shadow* of an argument in favor of the practice. We have read many recommendations of root-pruning, and have conversed with many
cultivators who were advocates of the practice. But when
we have summed up the whole matter, by inquiring for
the evidence, aside from an assumption—an *ipse dixit*—the
whole conclusion and evidence were resolved into this: "I
can not prove, positively, that root-pruning was beneficial;
but I am satisfied that the practice is advantageous." We
are not left in any such uncertainty with regard to the advantages of top-pruning. We have frequently read directions for severing such and such roots, which it would be
easy to show should not be pruned at all; and, if severed,
the tree must sustain a severe injury. The truth on this
subject is, that the instances in which an apple-tree needs
root-pruning are like angel's visits.

The operation of root-pruning is usually performed about
as skillfully as top-pruning would be, were the branches
thinned out by shooting repeated charges of large "grape-

shot" through the tree-tops with a cannon. In some instances, where the soil is so rich as to induce the branches to
continue to grow late in the summer, and even during the
months of autumn, if the growth is not checked by pinching
the spray in time to allow the cambium of the tender twigs
to solidify, the new growth will die during cold weather. Root-pruning will check that rampant growth; and if
the pruning should be severe, the flow of sap will be cut
off to such an extent that the young twigs can not be matured. If the flow of sap be cut off, the rampant growth
will be checked, frequently, to the serious injury of the
tree. Young shoots must be supplied with sap while the
cambium is solidifying. Hence, the true way to manage a
tree or bush, when it continues to grow rampantly, late in
the season, is to pinch the terminal buds instead of severing
the roots. This operation will stop the lateral and upward
extension of the branches, and the sap and cambium will at
once begin to develop the buds, thus inducing fruitfulness,
and promoting the maturity of the green wood and tender
branches, so that before cold weather every twig and shoot
will be fully ripened, and prepared to endure the rigors of
winter without injury. By pinching the terminal twigs
late in summer on some trees, and late in September on
others, the pruner will have complete control of the growth
and development of the branches. But let him undertake
to effect the same object by root-pruning, and there will be
more danger of injuring the trees than of securing the object in view. A tree should never be root-pruned except
by a person who has had extensive experience in the culture of trees. Let beginners beware of this unnatural and
dangerous process of mutilating the roots. If the terminal
shoots of a tree seem disposed to continue to grow after
the period has passed when all the wood should have been
fully ripened, take the pruner, Fig. 60, p. 133, and clip off

the ends of all those twigs that can not be reached with the hand. In some instances, twigs may commence growing again, and will require a second clipping. By treating a tree in this manner for one or two seasons, this rampant growth can be so controlled that the great flow of sap will promote abundant fruitfulness rather than surplus wood.

HUBBARDSTON'S NONSUCH.

A fine, large, early winter fruit, which originated in the town of Hubbardston, Mass., and is of first-rate quality. The tree is a vigorous grower, forming a handsome, branching head, and bears very large crops. It is worthy of extensive orchard culture.—Fruit large, roundish-oblong, much narrower near the eye. Skin smooth, striped with splashes and irregular, broken stripes of pale and bright red, which nearly cover a yellowish ground. The calyx open, and the stalk short, and in a russeted hollow; flesh yellow, juicy, and tender, with an agreeable mingling of sweetness and acidity in flavor; season, October to June.

CHAPTER VI.

GENERAL MANAGEMENT OF ORCHARDS.

Be mindful, when thou hast entombed the shoots,
With store of earth around to feed the roots;
With iron teeth of rakes and prongs, to move
The crusted earth and loosen it above.—VIRGIL.

THE duties of an orchardist are only begun when the young trees appear where they are expected to produce annual crops of fruit. We offer this suggestion for the purpose of waking up those "take-it-easy" cultivators, who have imbibed the erroneous impression that, after they have planted apple-seeds, and have thus produced trees as high as their shoulders, or have transplanted trees from the nursery, their labor and anxiety as to the future orchard are ended? There was never a more palpable delusion. Young apple-trees will require more care, at certain seasons of the year, than a flock of sheep; and during a large portion of the growing season, the orchardist must spend a great deal of time, and expend much labor in his orchard. If the soil is thoroughly cultivated, he must examine every tree frequently, to see if all the branches and twigs are growing correctly. At the same time, he must not neglect to wage an incessant war against noxious insects. All these duties must be attended to at the proper period. We can not have a *general time* of cultivating the soil, of pruning, or of combating noxious insects. The irksome duties in a young orchard which demand attention to-day, may need to be repeated to-morrow and the next day. The man who purposes to produce a profitable orchard of beautiful trees, which will yield bountiful crops of fruit for an age after he has passed away, must begin right, plant right, train right, and cultivate right. Then his reward will be as certain as the vicissitudes of the seasons.

Stirring the Soil in Orchards.—Many young orchards that are growing where the soil is thin, having a compact substratum beneath, are often *root-pruned* to their serious injury when the ground is ploughed. Young fruit-trees seldom have any roots to part with. Consequently, every rootlet that the plough severs tends to retard the growth of the tree. But where the soil is so porous that most of the roots strike deep, and spread out below the range of the plough, that implement may be employed for working the soil. None but a careful and intelligent teamster should be permitted to work around fruit-trees with any implement; and for several feet around every tree, the plough should not be permitted to run more than two inches deep.

The entire soil where an orchard is growing should be either mulched, or cultivated, or hoed over so frequently during the growing season, that all vegetation will be kept completely subdued. Indian corn, potatoes, turnips, carrots, or beans, may be cultivated between the rows every year for ten years. But the ground round about each tree, as far as the branches extend, should be left entirely free from vegetation. A few inches in depth of the surface should be kept stirred with hand-hoes, or horse-hoes, or harrows. Great care should be exercised in keeping all implements from wounding the bodies of young trees. A careless booby will frequently do many dollars' damage with the whiffletrees, plough, or harrow. A spade should not be used around fruit-trees to the injury of the roots. A spading-fork is better, as the tines will crowd the roots aside, seldom breaking even the small ones. Then, as the hard soil is broken up with fork-tines, removed from the roots, and returned to them thoroughly pulverized, all the little fibres will be brought into contact with different portions of soil that have not been exhausted of their fertility. Thus comparatively new earth will settle around the roots, so that in a short time the spongioles will begin to absorb plant-food. Now, if a spade be used, such a large proportion of the roots will be severed, that much of the source of plant-food will be cut off. Forking around fruit-trees is recommended only in certain instances. Mulching is better than hoeing and spading, or scarifying with a horse-hoe. When the soil is only a few inches deep, and the subsoil so compact that but few roots can enter it, a careless man with a spade will 'cut off more than half of all the roots, which are the main sources of nourishment; and the growth of the tree or plant will be retarded quite as much as if it had just been transplanted. When a spadeful of soil is filled with small rootlets and fibres, the spader had better be spending his time in idleness, than mutilating the roots of either ornamental or fruit trees.

Many an excellent orchard has been nearly ruined by directing some strong ploughman, who has been accustomed to plough new land, to rip up every square foot deeply. After fruit-trees have commenced bearing, the ground should not be touched with a plough within a distance of eight feet from them. If the soil requires renovating, top-dress and mulch it. If it is so porous that the roots will strike down readily below the reach of a common plough, as roots of trees always do where there is no compact substratum, there will be but little danger of mutilating the roots of fruit-trees. A ploughman should possess sufficient knowledge to judge correctly touching this subject. In many sections of country, the surface-soil, resting on a compact hard-pan, is so thin that a plough, if run ten inches deep, would cut off almost the entire system of roots.

A Thill Horse-hoe. — The best horse implement that I have ever met with for scarifying the soil in an orchard is represented by Fig. 64. With such a hoe the surface of

Fig. 64.

A thill horse-hoe with cast-steel teeth.

the ground can be scarified two or three inches in depth, much faster than a score of men can accomplish the same labor with hand-hoes, or with spading-forks and rakes.

As this implement is guided by thills and handles, the teeth can be run very close to the young trees, and be held to just skin the surface as it passes the trees. By this means the surface-roots of young trees will not be mutilated.

How to Plough around Trees.—If there is a grassy sod near the trees, it will be almost impossible to cut it all up with the plough. Before the ground is ploughed, a spade, or a bog-hoe, should be employed to cut up and turn over an area of at least four feet in diameter. Then there will be no necessity for running the plough too close to the trees. If the plough is drawn by oxen, great care must be exercised by the driver to keep the team from rushing astride of a tree that the oxen can bend to the ground. Well-trained oxen will often hook their under-jaw around the body of a small tree as they are passing it, and bend the top to the ground in an instant. They like to demolish trees that they can bend to the ground. That the plough-man may have complete control of his plough, there should always be a gauge-wheel beneath the forward end of the beam, so that the ploughman may lift his plough entirely out of the ground at pleasure. Also, when the plough approaches a tree, the ploughman should lift on the handles, gauging the depth of the furrow at pleasure, near the trees. If roots appear near the surface, let the plough run only two or three inches deep as it passes the trees. When horses and whiffletrees are used, the whiffletrees should not be more than twenty-two inches in length, and the driver should exercise great care to prevent the injurious contact of the whiffletrees with the tender trees. If there is no sod on the ground, it is an excellent practice to do the ploughing near the trees with one horse, or with two horses geared *ad tandem*—one forward of the other. So long as trees can be bent sideways, one person should be employed to bend the trees away from the team, and to elevate the outer end of the whiffletree while passing each tree. Otherwise, young trees will be seriously damaged by whiffle-trees. A slight touch with a whiffletree will frequently remove a piece of bark as large as a man's hand. Such wounds will be a great injury to the growth of young trees, unless they are covered immediately with a heavy coat of grafting-wax. To say nothing of the injury to fruit-trees by wounding the smooth and straight stem, it looks bad to see young trees stove up by unskillful and heedless ploughmen. We have in mind a young farmer who cultivated a large young orchard for more than ten years; and not a scar could be found on one of the trees. No one except himself was permitted to manage a team or to handle an implement near those trees.

Muzzles for Teams.—No team, whether ox, ass, or horse, should ever be permitted to come near young trees of any

kind until a good muzzle is secured over the mouth, as represented by Fig. 65. A pair of wire
muzzles may be obtained at most
hardware stores in large cities for
about twenty-five cents each, which
will be serviceable for an age, if
properly taken care of when not
in use. Muzzles are usually fastened to the bits of a horse's bridle,
and to the heads of oxen, by tying

Fig. 65.

A wire muzzle on the nose of a
horse.

a strong cord to one side of the muzzle, and passing it over
the top of the animal's head to the other side. A good
muzzle may be made in a few minutes with narrow strips
of firm leather. Every intelligent teamster knows so well
that he does not need to be told, that most teams will crop
off the entire growth of one season at a single bite, when
they are passing a tree; and nothing but good muzzles will
prevent such a casualty.

Harrowing Orchards.—Most kinds of harrows are apt to
operate roughly when drawn near young trees. Many times,
when the driver has made fair calculations for the implement to pass a tree, the opposite side will encounter some
obstruction, or hang in the hard soil, and thus be the means
of throwing the opposite wing of the implement so forcibly
against the body of the tree as to make a large wound.
This is especially true of such harrows as have bands of iron
near the ends of each arm. To avoid injuring trees, Mon-
roe's Rotary Harrow will be found an excellent implement
for harrowing orchards, with which it is almost impracticable to injure a young tree when the implement is drawn
past it, as the point of the harrow that might come in contact with the tree will remain stationary against the bark
until the implement has passed so far along that every part

will move away from the tree, leaving it uninjured.   If this harrow had no other points of superior excellence, orchardists would do well to keep one expressly for working among fruit-trees.   But it has been in use for so long a period, that tillers of the soil are familiar with its great superiority for all kinds of harrowing, and especially for harrowing in cereal grain.

Cultivating Young Orchards.—One of the most efficient implements for this purpose is represented by Fig. 66, the

Fig. 66.

Nishwitz's disk scarifier.

pulverizers of which consist of several sharp-edged circular disks about one foot in diameter, being concave on one side and convex on the other.   When the wheels or disks are cast, a round steel pin, about three-fourths of an inch in diameter, is inserted in the mould, thus furnishing a steel journal for each disk.   A bolt, with a nut at the upper end, is passed through a socket-standard, which holds the disks in their position.   When the scarifier is in use, the disks are set at any desired angle to the line of draught, and each disk thus pulverizes and turns over a narrow furrow-slice. The disks operate by cutting, lifting, and turning over a few inches in depth of the entire surface of the land. Wherever it has been thoroughly tested, this implement has given excellent satisfaction as a scarifier, or as an implement for covering seed-grain.   The wooden frame consists of two pieces of hard, tough timber, about two inches in thickness by seven or eight inches wide, held in position by the cross-bar, which is firmly bolted to the side pieces. By taking out the bolts which secure the cross-bar, the wings can be spread farther apart or brought nearer together, as represented by the cut, copied, by permission, from the catalogue of agricultural implements of the "Peekskill Plough Works," Peekskill, New York.

Advantage of Working the Surface.—When ground is ploughed in the former part of the growing season, and the surface is cultivated or worked with a gang-plough every two weeks during the summer, the substratum will be kept so moist that the roots of growing trees will spread with far greater rapidity than it would be possible for them to extend if the ground were more dry and hard. Every observing tiller of the soil knows that ground on which wheat, oats, or grass is growing, during hot and dry weather will often be as dry as ashes, while the same kind of soil by the side of it, which is being summer-fallowed, will be quite damp. The leaves of the growing crop pump up and evaporate the moisture in the ground. By summer-fallowing the ground where young apple-trees are growing, the entire plot is kept moist, and as soft as the ground was early in the spring. Of course soft and damp ground—not wet land —in hot weather, greatly promotes a healthy growth of the trees, so that in many instances young apple-trees will grow, where the soil is thoroughly worked, twice as rapidly as if a crop of grain or grass were cultivated on the same ground.

We have heard "gentlemen farmers" contend earnestly that a crop of grass, and even a heavy crop of weeds, will operate in a most favorable manner in dry and hot weather towards keeping the ground damp by the shade of the leaves and branches. If the growing plants did not pump up immense quantities of water, their reasoning would be sound. Frequent working of the soil in hot weather greatly augments its capacity to condense moisture from the atmosphere, and to retain it after it has been absorbed. Cultivation is essential to the thrifty growth of any apple-tree. The difference between the size of trees when cultivated, at the end of five years, and of those allowed to stand in grass, will be greatly in favor of the former. It is not the land we wish to improve, but the tree; it is not potatoes and beans we desire to raise, but to fit the soil in such manner that hereafter it will give food enough to the tree to enable it to raise large crops of apples. Every tree should be tilled like a hill of corn or potatoes. Yearly, as the roots extend, a wider space around the trunk should be cultivated. The whole ground should be spaded and hoed, mulched and manured, or scarified from row to row.

Ridging Fruit-orchards.—The aim should always be to keep the surface of the ground as level as a lawn. Apple-trees should never be "earthed up," as is sometimes done, by ploughing the ground in ridges, and turning the furrows at every ploughing towards the trees. J. W. Clarke, of Wisconsin, gave an account recently of the fatal injury to a young orchard on a farm which he had recently purchased. He says that, "the owner having left the State, the land was rented to three or four successive tenants, and each one took the *easiest* way to plough the ground; and, as the result proved, the surest to destroy the trees. At every ploughing, for five years, the ground was ridged up against the trees, so that when I took possession the stems were earthed up, on an average, eighteen inches, and some of them considerably more, above the depth they were set, and of course above that at which they stood in the nursery."

"The result was that, between 1857 and 1861, one-half the trees were half rotted through, *above* the general level of the ground, but *below* the tops of the ridges, which stupid laziness had formed; over two hundred fine young trees were rotted, broken off, and destroyed by the tenants in that heedless fashion."

The probability is, that most of those trees had been badly wounded with ploughs and harrows, which injured them more than the ridges. Nevertheless, it is an injurious practice to increase the depth of the soil near young trees to any considerable extent.

Mulching Material for Apple-trees. — Doubtless many persons have noticed how much more productive an apple-tree seemed to be, when the surface of the ground round about was covered with chips, or chip-dirt so thick as to kill vegetation. Tan-bark and sawdust will subserve the same purpose. If those persons who live near tan-works or saw-mills would get the tan-bark and sawdust, and mulch the trees every year, or every second year, and pile on both coal ashes and wood ashes, they would soon have abundance of choice fruits, as well as thrifty trees. There is too much of our woody material allowed to waste, which, if applied to lands on which trees are grown, either fruit or ornamental, would soon show its beneficial effects. It is folly to think of raising a bountiful crop of grass or grain near apple-trees, and to get an abundant crop of fine fruit at the same time. The mulching material can be applied in the winter, when laborers and teams have little to do. It will pay well to haul tan-bark, sawdust and planing-mill shavings two or more miles, when a large load can be carried, for the especial purpose of scattering it round about fruit-trees, unless trees are standing on a deep and fertile soil, so abundantly supplied with fruit-producing material, that nothing more is required. It will pay to spread coal ashes round about trees so abundantly that no grass or weeds can come up through the ashes. Any kind of straw, leaves, wild grass, sedge, or flags, may often be employed for

AUTUMN STRAWBERRY.

Fruit, medium, roundish, slightly conical, nearly the whole surface covered with bright red streaks on yellow ground; stalk, slender, about an inch long; basin, ribbed; flesh, yellow, very tender and juicy, with a fine sub-acid flavor. Tree, thrifty. Ripens early in autumn, and keeps well. Very productive; one of the best early autumn apples.

mulching young apple-trees. On light and sandy soil, it would be an excellent plan to spread two or three tons of clay round about each tree. Then, after the frost and rains of winter had thoroughly pulverized the clay, let it be forked into the sand. After which, apply a dressing of mulch. When mulch is applied in the spring of the year, let it be spread over the surface barely thick enough to prevent grass growing. A thin mulch will keep the soil moist, when without it the surface would be as dry as dust. Five dollars' worth of mulch will often save ten dollars' worth of labor in hoeing and watering plants. Many people remove all the weeds and grass to the street, when such things should be spread around growing vegetables or trees.

By many beginners, who do not understand how to apply mulch, a small heap of material is piled up close to the foot of the tree, without one moment's reflection as to the position of the roots that need the benefit. Now it may be laid down as a rule which is not far from being correct, that the length of the roots which radiate on all sides from the base of an apple-tree is about equal to the height of the tree itself. If, for example, a young fruit-tree is ten feet high, then we may infer that the roots form a circle about twenty feet in diameter, the base of the tree constituting the centre. Over this great surface, the finer and more inconspicuous roots form a net-work of fibres; and to derive full benefit from manure, cultivation, and mulching, a broad space of ground must be covered with manure and mulch.

Wood Ashes for Apple-trees.—Many of my readers, doubtless, can recollect instances where a fruit-tree has stood near a large heap of leached ashes; and sometimes the ashes have been thrown out of the leach for several years, so that every noxious weed and grass has been killed by the liberal top-dressing of wood ashes. But the fruit that was produced on the tree near the ashes was always plump and smooth. I well remember, when a small lad, that there were two peach-trees near my father's leach, around which the leached ashes had destroyed all vegetation; and the old trees seemed to cling to life with the desperation of a drowning man. But the fruit was always large, and very smooth and luscious, as long as there was sufficient vitality in the last twig to produce a fruit-bud. This fact assured me that fruit-trees require wood ashes. Consequently, after I commenced operations on my own farm, every bushel of ashes was spread around fruit-trees; and many loads were hauled five miles to be scattered around fruit-trees. The result was, that in after years, when fruit-trees in all that vicinity yielded scarcely a family supply of fruit, my trees, that had been top-dressed with ashes, bore bountiful crops of smooth fruit. I have in mind several old trees which would yield annual crops of small, gnarly, knotty, rusty, one-sided fruit of little value, until after I had made a portable pen, about eighteen feet square, around each tree, in which two fattening shotes were kept, about two weeks per tree, until they had rooted up every inch of the turf and destroyed the grass. Then several bushels of unleached wood ashes were scattered around those trees, as far as the roots extended. The result was, bountiful crops of plump and smooth fruit every succeeding season; and I meet with farmers every year who tell me that their apple-trees and Vergalieu pear-trees which stand near their leach, and that have been top-dressed with ashes, yield bountiful crops of luscious fruit; while the same kind of trees near by, that have not been supplied with ashes, seem to be failing; and the fruit on them is rusty, one-sided, and very inferior. I once had an Early Harvest apple-tree that always yielded small, knotty, and cracked fruit, until a few bushels of unleached ashes were spread about it, after which the fruit was superb.

Bones for Apple-trees. — There is no more valuable fertilizer for apple-trees on most soils than bones, and the flesh of animals and fish. It has been recorded by reliable authority that near the graves of Roger Williams, the founder of Rhode Island, and his wife, there stood a venerable apple-tree which had sent two of its roots into the graves of Mr. and Mrs. Williams. The larger root had pushed its way through the earth till it reached the precise spot occupied by the skull of Roger Williams. There, making a turn, as if going round the skull, it followed the direction of the backbone to the hips. Here it divided into two branches, sending one of them along each leg to the heel, where both turned upward to the toes. One of these roots formed a slight crook at the knee, which made the whole bear close resemblance to a human form. There were the graves, emptied of every particle of human dust. Not a trace of any thing was left. There stood the guilty " apple-tree," as it was said at the time, caught in the very act of " robbing the grave." The fact proved conclusively that bones, even of human beings, are an excellent fertilizer for fruit-trees; and the fact must be admitted that the organic matter of Roger Williams had been transmitted into the apple-tree; it had passed into the woody fibre, and was capable of propelling a steam-engine; it had bloomed in the apple-blossoms, and had become pleasant to the eye; and more, it had gone into the fruit from year to year, so that the question might be asked, Who ate Roger Williams?

It is known to chemists that all flesh, and the gelatinous matter giving consistency to the bones, are resolved into carbonic acid gas, water, and air, during decomposition, while the solid lime-dust usually remains. But in this case, even the phosphate of lime of the bones of both the graves was all gone. In the same manner, all kinds of bones, when buried near growing vines and trees, will be literally devoured by the hungry roots; and the once beautiful forms that lived and moved will be transformed into luscious food, to maintain others who may exist hereafter.

Scraping the Bodies of Apple-trees.—This practice is repeatedly recommended as an important means of producing a fine crop of fruit. But we can conceive of no practical advantage in scraping old trees so thoroughly as to remove all the shaggy and dead bark, except to destroy insects, their eggs, and their refuges. The rough and shaggy bark of an apple-tree will be found advantageous to the health and productiveness of a tree, as the covering is provided by nature for their protection against the adverse influences of climate, and for maintaining in even balance that degree of warmth most favorable to the healthy exercise of their vital functions; and perhaps the fact is not sufficiently realized how admirably suited this external covering is for such a purpose. Wood, in the direction of across the grain, is a non-conductor of heat in an eminent degree; and the outer layers of bark are even greatly more so. Hence we would never advocate the practice of scraping the bodies of fruit-trees simply to promote their fruitfulness. The practice promotes no such end. When the bodies of trees *are* scraped, the denuded parts should be protected with a coat of pitch and tallow. (See p. 159 for a cut of a tree-scraper.) The chief reason urged in support of the practice is, that it lightens the trees of a load of dead and useless matter which can not be of further benefit; but, on the contrary, when allowed to remain, it is only a harbor for insect enemies, and for the growth of moss and lichens. It is doubtless true that the depredations of insects have been lessened by the process. But as for lichens, we are at a loss to understand how they can be hurtful to trees; for, although growing upon them, they live exclusively upon food drawn from the atmosphere. It is not claimed that trees derive any benefit from those epiphytes, unless in bleak and exposed situations they supplement the bark in affording protection. But when found in excess upon apple or pear trees of immature age, they may be accepted as a sign that, from some cause or other, those trees are not in such satisfactory condition as could be desired. Want of thriftiness in trees favors the growth of lichens; and this may be induced by a variety of causes, either singly or in combination, such as impoverishment of the soil, want of drainage, or exposure to the rigors of a severe and variable climate.

The long strips of dead bark hanging from the bodies of large hickory-trees are of more service than injury. In no supposable case need the pomologist hope to better his prospects by the use of a scraping-knife; for, from whatever cause trees cease to increase perceptibly in diameter, outside appliances can no more impart fresh vigor to them, than the under-fed and over-driven roadster, all of whose bones stick out, can be made sleek and fat by the use of the curry-comb, so long as generous feeding, careful driving, and a good stable are ignored. In the vegetable kingdom, the fitting of means to ends is so close and intimate as not to be safely disturbed; while, at the same time, they seem to

Fig. 67.

Tree-scraper.

indicate that the practice we oppose is out of harmony with the plans and purposes of nature. Hence, when we remove the dead bark of an apple-tree, we should first understand what the effect is to be on the tree. We would never scrape or shave off the dead bark without being able to tell *how* the operation will promote fruitfulness. So long as a tree continues to yield fair crops of apples, the bark should not be tampered with. Fig. 67 represents a scraper of convenient form, which consists of a triangular piece of steel, about three inches from corner to corner, with the edges ground or filed sharp for scraping. A stiff broad hoe may be employed instead of such a scraper.

The editor of the *Practical Farmer*, Philadelphia, says:

"Some over-wise people have an idea that when a tree gets mossy and bark-bound—the latter but another term for the want of growth and weakness consequent upon neglected cultivation—it is only necessary to slit the bark up and down the stem with a jackknife, and it will at once spread out and grow. This is sheer nonsense. Dig about and cultivate the roots, and the bark will take care of itself, with a scraping off of the moss, and a washing of the stem with ley or soap-suds, or chamber-slops, which last is quite as good. The increased flow of the sap, induced by a liberal feeding of the roots, will do its own bursting of the 'hide-bound' 'bark, which is simply its enfeebled condition as a consequence of its poverty of root. No one thinks of turning out a bony, half-starved calf in the spring, into the clover-field, with the skin on its sides all split through with a knife, in order to add to its growth. And this last proposition is quite as sensible and philosophical as the other. Nature takes care of itself in these particulars. Sap in plenty is what the blood is to animals. Its vigorous flow reaches every part of its composition, and gives to each its proper play and function. We can show frequent instances of a decrepit shrivelled branch, by the throwing open and manuring of the roots, and a thorough pruning of the whole top, increasing from an inch to two inches diameter in a single season, and without assistance, as it grew, bursting and throwing off its old contracted bark as freely as the growth of a vigorous asparagus-shoot would develop itself during a warm shower in May. Such nostrums are only the invention of the head to excuse the laziness of the hands."

Girdling and Peeling Trees.—During the former part of summer, and sometimes as late as the first of August, the bodies of apple-trees may often be denuded of every particle of bark, from the ground to the branches, with excellent effect, *in some instances,* on the productiveness of the tree. But such an operation must not be performed too early, nor too late in the growing season. There must be a liberal supply of cambium, or new wood in a semi-fluid condition, between the old bark and the wood. Then, if the old bark be removed, the cambium will form a new coating of bark, instead of another concentric ring of new wood. We have frequently seen apple-trees completely stripped of the bark, clear up among the limbs. On such trees, before the end of the growing season, a smooth and thick bark had been formed, wherever the old bark had been disturbed.

The object in view is, to promote fruitfulness. In many instances, "*ringing*" the body of a tree, or the main branches, by removing a strip of bark quite around them, half an inch or an inch wide, will have the desired effect. In some instances, a sharp knife has simply been run around the limb, cutting through the bark, without removing any portion. Ringing of branches will sometimes promote early maturity. We have often seen branches loaded with mature fruit, more than a week previous to the time of ripening of the same kind of fruit on other branches of the same tree. Then, again, we have ringed branches with our own knife, and no effect whatever could be perceived in the productiveness of the branches, or early maturity of the fruit.

The philosophy of ringing branches, or peeling off the bark to promote fructification, consists in this: that the operation retards the growth of the branches, which tends to promote fruitfulness. But some of our most reliable pomologists contend that all such operations are highly injurious to the trees, because they are unnatural. We would not advise the practice, except when the bark is old and the trees "bark-bound."

Thinning out young Fruit.—This is a practice that has numerous advocates, and only a limited number to practise it. Ever since we commenced the cultivation of fruit-trees and grape-vines, we have not failed to thin out the fruit— especially of young trees—with as much care as the same trees were pruned. When young apple-trees or pear-trees have been covered with blossoms, and a sufficient number of specimens have set to make several quarts of fruit, our practice has ever been to clip off all but three or four specimens, particularly if the branches of the trees were growing slowly. It is exceedingly detrimental to young trees to be allowed to mature much fruit while they are growing vigorously.

The most convenient instrument for thinning out fruit is a small pair of sharp shears, with which to cut the stems. One man will be able to thin out the fruit on a large number of trees in a day, if he can handle his shears with facility. When the young fruit hangs in clusters, always cut out the poorest specimens. Do not be chicken-hearted about such a job, although it appears terribly destructive to your crop of fruit. Faint-hearted women will hold their breath at the sight, and even venture to scream as the young fruit falls. But act the part of a wise and skillful pomologist, and thin out all the clusters. The crop will certainly be more abundant, and more valuable also.

Hon. M. P. Wilder, President of the American Pomological Society, says that

"One of the most important lessons which experience has taught us, is the necessity of thinning our crops of fruit; and no operation, in the whole round of fruit culture, has been so much neglected. It is not strange that the young cultivator, delighted with a fine show of large pears on a young tree, or an abundant crop of grapes on a young vine, should, in the pride of his heart—and not knowing the impossibility of the tree or vine bringing them all to maturity without a tax on its vital powers, of which the effect may be felt for years—permit them to remain. It is true that the labor is great, but so is the profit; and oftentimes it happens that the labor of thinning a crop makes all the difference between unsalableness and a high price in the market. Not unfrequently a pear-tree will set so much fruit that it can not bring any part of the crop to a size which will render it salable in a crowded market, when, if one-half, or even a larger part, of the fruit had been removed, the remaining specimens would have sold quickly at the top prices of the market. One of our farmers, near Boston, always thins his fruit; another, adjoining his orchard, neglects it. The location and treatment of these two orchards, in other respects, are much the same; but the former realizes for his crop of Baldwin apples about four dollars and a half per barrel, while the latter, standing by his side in the market, receives less than three dollars and a half for his. The case is still stronger with the pear, which, growing on smaller trees, is more easily thinned; and the prices obtained for the fruit afford a better remuneration for the labor of thinning. While those properly thinned and cared for will command four dollars per bushel, those of the common run will not bring more than two dollars; and this rule applies not only to fruits, but to all vegetable productions. Every one has observed that the overbearing of a fruit-tree one year is likely to result in barrenness the next. Hence the necessity of thinning our fruits, so as to avoid exhaustion of the tree, and to keep up a regular succession of good fruit. Even the Baldwin apple, which, from its great productiveness, bears only on alternate years, we think, might, by thinning, be made to bear annual crops. Not merely the form, but the color, is improved by thinning; for without sunlight fruit can never attain perfect color. When apples are crowded in clusters, they are particularly liable to be attacked by disease; and therefore the necessity, if we wish perfect specimens, for removing a part, so that no two fruits shall touch each other. This necessity is especially strong in the case of the peach and plum, and early apples, where rot is liable to be communicated by contact. These should be so severely thinned, when young, as to make it certain that they will not touch each other when fully grown."

Making Letters on Growing Apples.—It is a curious fact, which many pomologists will doubtless be glad to know, that a person may produce his name in beautiful letters in the skin of an apple, thus: Cut paper letters out of a newspaper, say one-fourth of an inch square, spread a coat of mucilage on one side of each letter, and stick the letters on the surface of an apple that is growing in the sunlight. The letters should be put on before the apples have turned red. In ten to twenty days the apple will be red, except beneath the letters, which will be of a light color. After the apple is done growing, the paper may be removed, and the letters or figures will remain as permanent as the color of the fruit. Many a person would be delighted with the idea of sending a large and beautiful apple to a respected friend or lover, with his or her name in beautiful *plano-relievo*, formed on the smooth surface of a ruby apple by Dame Nature's own delicate touch, in shades and tints of unrivalled beauty. Another manner of recording a name and date is to simply cut through the skin of an apple in the shape of the letters. A name may also be made by punching the skin with a bodkin in the form of letters. The wounds will heal so that the scars will form the letters of the name intended.

Preparation of Apple-trees for Winter.—In some instances, when an apple-tree is standing where every thing favors a luxuriant growth, the branches will keep green until cold weather and frost has destroyed the foliage. A strong, perfect, and healthful animal will endure the rigors of our Northern winters with far less injury than a poor, weak, and half-grown beast. The same principle will be found to hold good in the management of apple-trees. If their shoots and buds are full and plump, and well supplied with healthful material contributed from clean, healthy leaves, the chemical movements which attend growth will assist greatly in maintaining the tree against cold, by the heat which is developed. In a thin, weakly tree, this force is wanting. Hence the eminent importance of watching the growth of young trees early in autumn, when the growth is rampant, for the purpose of clipping the extremities of luxuriant branches, to hasten the maturity of the new wood. Untold numbers of valuable fruit-trees at the West and North-west have been lost during cold weather, from no other cause than the one here alluded to — luxuriant growth in late autumn. In many instances, trees have grown but little

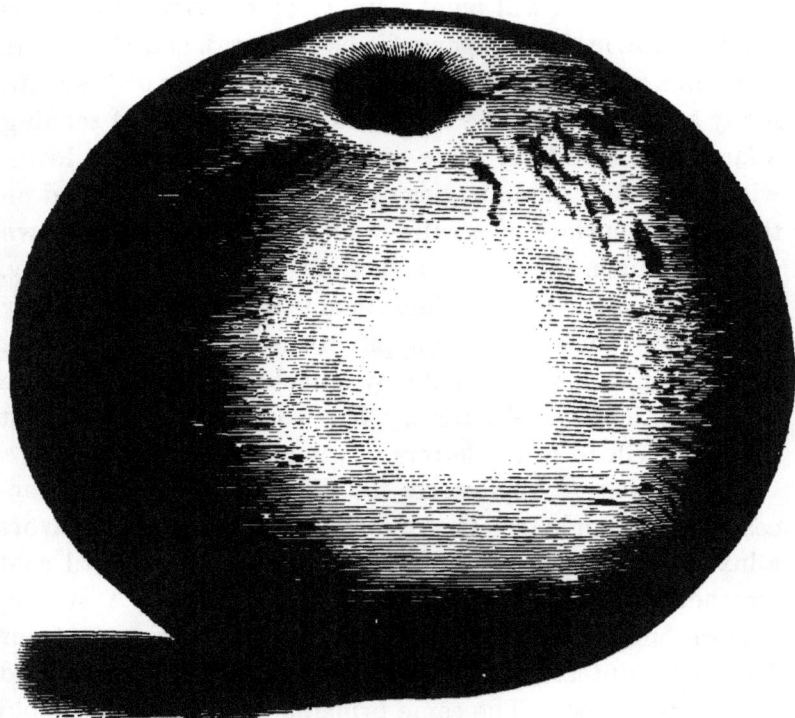

RHODE ISLAND GREENING.

Synonyms. — Burlington Greening, Jersey Greening, Russine, Grünling von Rhode Island, and Bell Dubois. The fruit is large and roundish, a little flattened, quite regular, but often obscurely ribbed, dark green, becoming greenish-yellow when ripe, sometimes showing a dull blush near the stem. The flesh is yellow, fine-grained, tender, crisp, with an abundance of rich, slightly aromatic, lively acid juice. For market and for culinary purposes, it has few superiors. Season from November to April. The tree is vigorous, and usually hardy. At the South the fruit is said to drop too early in autumn to be a profitable apple.

during the summer, on account of drought. When the autumn months have been exceedingly favorable for the growth of all kinds of trees, and when young branches have continued to push out until freezing weather has put an end to all vegetation, if trees have not been prepared for winter, of course many of them must be utterly destroyed by the cold.

Another thing in preparing apple-trees to endure the winter without injury—in addition to the foregoing—is a mulch or covering for the surface, with some vegetable material in a state of decomposition, which will shelter the roots and impart warmth, evolved by its slow combustion; and the more valuable this mulch will be if nitrogenous matter is included, as in stable manure. This application is of vastly more benefit, when applied in the autumn, than if left till spring, not only on account of the shelter it affords, but because of its advancing a strong growth early in the spring, which becomes well ripened before winter; whereas, manure applied in the spring, especially if raw, often does not become effective until late in the season, when the wood should be ripening instead of growing. A ripe, well-varnished and finished coat of bark is to the tree what the coat of hair is to the animal; and the effect of a small break or rent in it shows how important its perfect condition is, especially that of its outer skin or epidermis. It must be remembered that the dark only ripens well in full light. Trees, therefore, must have their wood both well fed and thoroughly ripened, or they will not possess sufficient vitality to endure the extreme cold of our Northern winters.

Certain writers at the present time are advocating the efficacy of *root-pruning* as the most reliable way to promote hardiness during the winter. But it will be found that by employing such an instrument as is illustrated on p. 133 to clip the terminal shoots, and thus induce the new wood to harden before cold weather, apple-trees will endure the cold more satisfactorily than if they had been root-pruned.

Protecting Apple-trees from Stock and other Animals.—

> If cattle and horses may enter to crop,
> Young trees are in danger of losing the top.—EDWARDS.

Horses and colts will nip off the branches and gnaw off the bark of young trees; horned cattle will browse the tender twigs, hook the tops, and break the stems down by rubbing unceremoniously against the bodies of young trees; and sheep, calves, and goats will frequently gnaw off all the bark within their reach. Horses, mules, and neat-cattle should never be allowed to run loosely in a young orchard. Sheep may run in an orchard if the bodies of the trees are first protected. The most convenient manner of protecting from sheep, goats, calves, rabbits, and mice, is to procure strong, coarse paper, or some coarse, cheap canvas; cut it in strips of the right breadth to wrap around the tree, and lap not less than two inches. If a piece is passed twice around, it will do no harm. Then secure the edge by driving large tack-nails — carpet-tacks — into the tree. Then apply a coat of coal-tar to the paper or cloth. Such covering should be removed during the summer. We have never heard of a protection that is equal to this for cheapness and efficiency. Such material may be taken off the fore part of every summer, and kept till the leaves have fallen. Small nails driven into a tree will not injure it. In many localities, when snow drifts deep around trees, rabbits will run on the surface and gnaw the bark, if not protected

Fig. 68.

Protected from mice, rabbits, sheep, and goats.

five or six feet from the ground. If coal-tar be applied to the bodies of young trees, there is danger that the poisonous material in it will injure the bark. I once applied coal-tar to a valuable apple-tree ten years old; and in less than eight months every root and branch was killed by the tar.

Fig. 68 represents a young tree having the stem wrapped with tarred paper, and a mound of earth cast up around it, to turn away field-mice. When a mound of earth is placed around a tree, the tarred paper need extend only two or three inches below the top of the mound.

Trees blown over.—A furious wind frequently turns a large apple-tree over, or blows it so nearly up by the roots that the body stands at an angle of forty-five degrees. If a tree be a valuable one, it may be straightened up at a small expense. If the subsoil is compact where the tree stood, let it be dug up and thoroughly pulverized, the deeper the better. Then hitch a strong rope or chain to another tree or post, and one end to the tree-top, having a system of tackles between the tree to be set up and the anchor-post. Let the rope pass over the top of two poles, bolted together in the form of a letter A, standing near the tree to be set up. When all things are ready, hitch a team to the slack-rope of the tackles, and set the tree up. Let the top be held erect until it is secured by four strong guy-wires, extending from the top to strong stakes of durable timber set firmly, three feet deep, in solid ground. After two years, if the soil is not too wet and compact, the new roots will have taken such a firm hold that a hurricane could not blow the tree over.

Manuring Orchards.— The land where some orchards are growing needs to be well manured before the trees are set out; and the surface should be top-dressed every two years. Any thing and every thing may be spread on the surface. Moderate doses of lime, more generous ones of wood ashes, are always profitable. Phosphates of all kinds are useful if buried beneath the surface. Even Peruvian guano, if applied at all, should be slightly dug in, late in the fall, so as to become thoroughly divided by winter rains. Good barn-yard manure spread over the surface of the ground is excellent. But coarse organic manure should not be put in contact with the roots of young trees. The more offal from the slaughter-house one has to put around apple-trees, the more scrapings of the manure-yards, the more chip-dust, the more fish, flesh, and bones of dead animals, the more of leather-shavings and hide-clippings, horn piths and hoofs, hair and old plaster, and all such articles, the better it will be for the growing trees. Apple-pomace, spread say one inch thick around a large tree, will be found a good manure, especially for old trees. If applied too bountifully, pomace will kill all vegetation, and the roots of the apple-trees. My father was accustomed to spread pomace about three inches thick over Canada thistles and elder bushes; and that thin top-dressing would destroy every. root of every plant—no matter what it was—clear down through the subsoil. Of course, if pomace be applied too bountifully, it will kill the roots of apple-trees.

Garbage for Apple-trees.—Many of our most successful pomologists believe that there is no danger of applying too much manure to bearing trees. Their fruit shows that such a theory is correct, provided fertilizing material of the required kind is employed. Many families in cities, villages, and even in the country, cast away enough fertilizing material in one year to equal the manurial value of several tons of the best Peruvian guano, which is sold to tillers of the soil for about eighty dollars per ton. Many a family throws out at the back door a sufficient quantity of garbage from the kitchen—consisting of apple-peelings, potato-peelings, bits of meat and bread, fragments of bones, feathers of fowls, and soap-suds—to produce all the fruit or vegetables that the same family would consume, if such material were properly applied to the soil. Many families cast into the garbage-barrel more than five pounds of pieces of bread, meat, and bones every day, every pound of which is more valuable as a fertilizer of the soil than a pound of guano. Other families waste several barrels of soap annually, every gallon of which is of more value to spread around fruit-trees than a pound of Peruvian guano. After a box of hard soap, or a barrel of soft soap, has been dissolved in the wash-bowl or wash-tub, its manurial value still remains in the water; and the fertilizing material is in a far better condition to feed growing plants and fruit-trees than before the soap was dissolved. Every barrel of good soap is worth from five to six dollars to spread around grape-vines or fruit-trees of any kind, as soap is the choicest quality of fertilizing material for bearing trees, and even for flower-beds. The more we can apply, the more productive the trees and the soil will be.

The great practical point, then, will be *how to save it.* In our own kitchen a garbage-pail is provided, into which egg-shells, apple-peelings, fragments of beef, mutton, fish-bones, and every thing of this character, is thrown; and as often as the pail is half filled, the contents are buried around our trees or grape-vines. The soap-suds are all carried into the yard, when the sun does not shine, and poured around trees, vines, flowers, or vegetables. It would be an excellent practice to provide a large tierce, molasses-hogshead, or water-tight box, place it on a stone-boat, like Fig. 69, or on a pair of low runners, and let it stand in a hollow a few rods from the kitchen, where it could receive all the dish-water and soap-suds. As soon as it is full, let a team be hitched to the

Fig. 69.

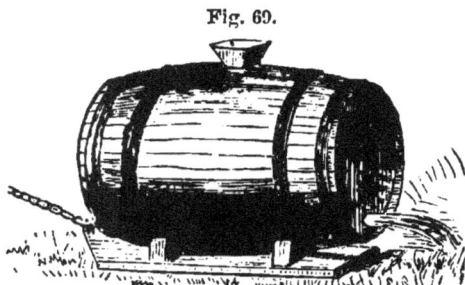

A liquid fertilizer apparatus.

sled, and the whole drawn to the fruit-yard, or to a lawn, where the contents could be let out on the ground in a few minutes by opening a valve, or withdrawing a large plug near the lower side. Whoever will practise this mode of

8

saving and applying valuable fertilizing material for a few years, will see a surprising improvement in his bearing trees, grape-vines or grass-land, where such substances were distributed. Better manure can not be desired. Besides this, the cost of it is comparatively nothing.

Under-draining old Orchards.—I have met with a large number of old orchards in New England, New York, Ohio, and other States, many of the trees of which do not yield a crop of any value, and never have, on account of the excess of water in the soil. And yet the proprietors of those trees are looking to some climatic influence, or to some "east-wind," or to a degeneracy in the variety, as the cause, beyond man's control, to which the failure of fruit must be attributed, when the cause is purely *local*, and might have been removed years ago. Old apple-trees will not flourish where there is an excess of water. In many instances, we have observed that apple-trees standing where the soil was kept thoroughly charged with water late in autumn, during the winter, and for two months in the spring, would be so seriously injured that they could not recover during the summer. When the pores of the soil are all filled with water, week after week, instead of warm air, an apple-tree can not thrive luxuriantly, nor yield satisfactory crops of fruit, until the land has been under-drained. (See p. 61.)

Management of Dying Trees.—It would be folly and labor lost to attempt to rejuvenate an old, dying apple-tree. There is a limit to the duration of apple-trees, as well as to human life, so that, after an apple-tree has been dropping a branch here and there for several years, and the body has begun to decay, we can not make a profitable tree of it, any more than an old, blind, lame, and foundered horse can be rejuvenated. The very best way to treat such trees is to dig them up, and start another tree of some valuable variety where the old one stood. Direct the digger to excavate, with pick and shovel, all around the old tree, three feet deep, and seven feet in diameter, cutting off the roots as he digs. Any experienced digger will sever an old tree in two hours, so that a team can roll it away. Let such work be done in the winter, or early in the spring. Then the ground will be in readiness for a young tree. As the old roots decay, the new roots will fill their places. Should there be a lack of apple-producing material in the soil, it should be supplied. Many people will cling to an old apple-tree so long as it will yield only half a dozen poor apples, when a young tree would produce several bushels of superb fruit.

Forming the Tops of bearing Apple-trees.—Fig. 70 represents a model dwarf apple-tree having a conical top. (See definition of *Pyramidal* in Glossary.) In order to produce a true top, lash a small pole to the central stem of the tree, as represented, then let a line extend from the top of the pole to the end of a *sway-bar*, one end of which is scolloped out to fit the body of the tree. Let the sway-bar be of sufficient length to carry the line beyond all the branches. Then, with the pruner (see Fig. 60), clip off the ends of all the branches that extend beyond the line. If it is desirable to form a round head, the line will be a reliable guide, by clipping off the ends within a given distance of the line, on every side, as represented by the diagram. A small block tacked to the tree beneath the end of the sway-bar will keep it from dropping to the ground. Such a device will be found exceedingly convenient when forming tree-tops of any other style.

How to Spread the Tops of Trees.—It seems to be a habit with certain trees to send their branches more erect than in a horizontal direction. This is particularly true of the Gravenstein apple and Bartlett pear-trees, and many other varieties. It is decidedly objectionable to have all the limbs

Fig. 70.

A pruning-gauge, to aid in giving the top a conical form.

shoot upward, forming a top so close that a person can not ascend between the limbs to pluck the fruit. Moreover, the fruit will be fairer, larger, and better, if the branches grow at a greater distance from each other, so as to let in the light of the sun and admit of a more free circulation of air through the dense foliage and thickly-set fruit. When there is ample space between the trees, fasten horizontal spars of wood to the main stem of the tree-top, and push the limbs outward each way from the middle of the top, and tie them in the desired position, with soft strings, to the ends of the horizontal spars. A more convenient way will be to pass a narrow strip of leather around a branch and *nail* it to the wood. The limbs may be fastened in the desired position by such means, and kept there for one season, when they will remain spread out, thus rendering it more convenient to move around in the tree-top when pruning, thinning the fruit, or plucking it. In some instances, only a few of the boughs grow erect on one side of the tree, which may be thrust outward and secured by short spars of wood on one side of the middle of the tree-top. If no animals are allowed to run in the orchard, a small wire may be secured to a limb, and fastened to a stake driven in the ground.

When limbs of trees have been bent down too low by any means, if the central stem is sufficiently stiff, such limbs may be elevated and wired-up for one season, after which they will maintain their position. Young trees require much attention when growing, to make every branch stand in the most desirable direction.

Grass in Orchards.—One pomologist will recommend the culture of red clover or grass, where a young orchard is growing. Another will denounce the practice of permitting grass to grow among trees of any kind. The advocate for clean cultivation argues that grass draws moisture from the soil faster than rains restore it; that the soil, not being stirred, becomes closely packed. The roots of the trees also fail to get a good supply of moisture when it is most wanted. The tree is exhausted in its effort to mature the crop. Besides this, if grass is allowed to grow round about fruit-trees, the roots are deprived of the best kind of a mulch—a mellow surface, frequently worked. It is a fact which can not be refuted, that heavy clay loams, which have a great capacity to absorb moisture, can be kept moist by frequent stirring of the surface during very dry weather, when a thin dressing of mulch would fail to keep the ground moist. But some soils are too porous to be kept moist by frequent stirring.

Well, what shall a beginner do, when authorities disagree? Adopt the system of management in which they *will* agree. When young fruit-trees standing in grass ground send out luxuriant shoots at the end of every branch, over one foot long annually, the trees are growing fast enough; and it would not be advisable to cultivate the surface, as the trees might be stimulated to produce a growth too luxuriant for the hardiness of the young branches. On the contrary, if the entire soil is kept clean and mellow through the growing season, and the annual growth of the branches does not exceed one foot in length upward and laterally, it is a certain indication that every tree needs a few bushels of ashes, a bushel or two of lime, or a dressing of clay, or sand, or muck, or barn-yard manure, or the soil needs underdraining. Grass would be ruinous to such trees. When it is desirable to keep the ground in grass, spread a bountiful top-dressing, and mulch round about every tree. There is an excellent influence exerted by the character of a thin, mellow surface on the roots of a growing tree. These roots may be at the first only *half a foot* below; yet daily stirring but an inch of the surface may quadruple the growth of the length of every branch.

The Twig Blight.—In many localities of the West, and occasionally of the East, this scourge of apple-orchards has appeared. In some instances, it has scarcely attracted attention. The true cause of the blight has not yet been discovered. Most pomologists conjecture and guess that some noxious insect is the cause of the great injury sustained by the trees. Still, repeated examinations with magnifying-glasses have revealed no evidences of the work of insects, as no stings, larvæ, or eggs have been discovered.

The most satisfactory explanation is the rather tantalizing one attributing the death of the twigs to the invasion of minute parasitic plants—fungi. Then, if fungi may be relied on as the true scourge, we have a remedy, which consists of thorough cultivation of the soil with scarifiers, a liberal top-dressing with lime and wood ashes, and a proper preparation of the twigs for cold weather, by pinching the terminal buds or clipping the luxuriant branches before the growing season has ended, so that the new wood may be thoroughly matured before cold weather.

Swine confined near Fruit-trees.—When swine are confined in a pen around fruit-trees, they should be watched closely, lest they strip all the bark from the trees. More than this, the pen should not be kept around one tree over ten to fifteen successive days. Fattening swine that do not receive a supply of grass, or fruit, or vegetables, will often devour every mouthful of bark that they can reach, whether it is found on the roots or the body of a tree. Strips of boards may be nailed firmly to a tree to protect the bark from swine. We have often heard it remarked by old men, that "it is sure death to a fruit-tree if swine are yarded around, and allowed to root much about it, and to sleep near the body of it."

That this is not *always* the effect of yarding swine around fruit-trees, will appear from what we have to state

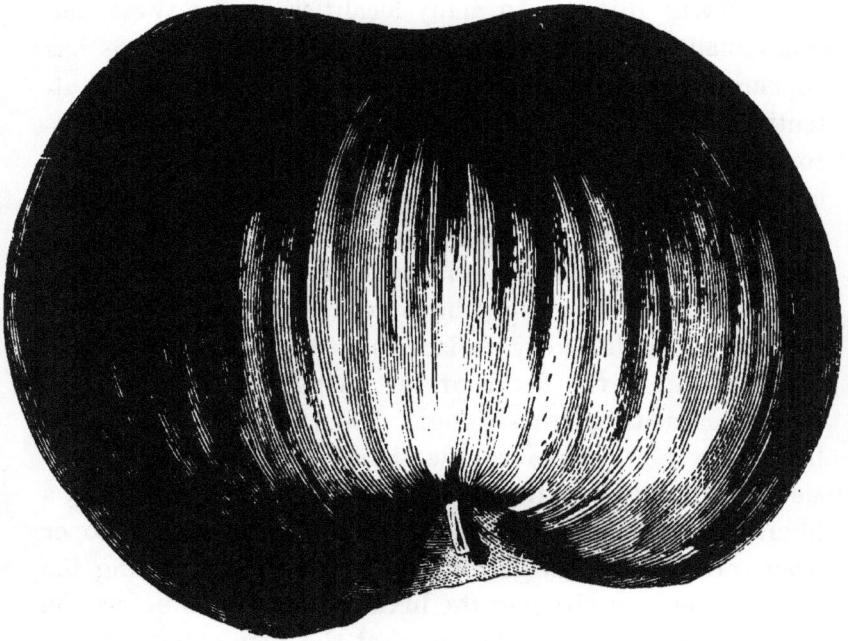

MYERS'S NONPAREIL.

Synonyms.—Ohio Nonpareil, Cattall Apple. *Fruit.*—Size, large; form, round-ish flattened; color, red and yellow marbled and splashed, and with many scat-tered gray dots; stem, short to medium, small; cavity, regular, open; calyx, par-tially open; basin, medium depth, smooth and regular; flesh, yellowish white, juicy, rich, tender, mild, sub-acid; core, regular, partially open; seeds abundant, plump, and full; season, last of September to early December. In many locali-ties, this valuable apple is highly esteemed both as a market variety and as a su-perb fruit for culinary purposes. In other places, it does not succeed so satisfac-torily as is desirable.

respecting a pear-tree, and a cherry tree, with which we once experimented. The pear-tree was upward of twenty years old, and had never produced any ripe fruit. It was usually well loaded with young fruit, much of which would swell to the size of a large hen's egg, and then would be-come knotty, full of cracks, wilted, and would all fall to the ground, long before it was time for any of it to ripen. Neither pruning nor manuring appeared to have any amel-iorating effect. We often examined scores of the fallen fruit, in order to ascertain whether or not the curculio, or some other marauder, were not the cause of such an untimely casting of the fruit; but not a vestige of an insect could be discovered. With no expectation of ever gathering any ripe fruit from that tree, we determined to test what was looked upon as one of the whims of our illustrious progenitors; and, accordingly, two shotes were yarded beneath the tree, in an inclosure about one rod square. They were kept there about two months. Their bed was close to the body of the tree, under a few loose boards. In this yard, holes were made with a crowbar, into which kernels of corn were dropped. The whole ground was rooted over and over, to the depth of eight to twelve inches; and many of the roots were torn up. This was in August and September. The next season, instead of seeing a *dead* tree—the result of recklessness, as was prognosticated —every branch was well filled with luscious fruit, and those trees continued to yield bountiful crops every season, as long as we were in possession of them. The cherry-tree was a few rods from the pear-tree, and was literally loaded with fruit every season. But previous to that treatment it never ripened. Some of it would rot. Some would wilt and dry up. Some would be covered with black knots, and some would become almost ripe, and then drop to the ground; no traces of insects could be found. The *hog remedy* was applied *thoroughly,* and every season since it has borne a large supply of as good cherries as ever birds picked.

From these facts it would appear that, if yarding swine about fruit-trees is *generally* injurious, here is an exception. If swine were permitted to sleep close to the body of a tree, and to root about incessantly for a whole season, we are not prepared to say that the effect might not prove fatal. There can be but little doubt, when fruit-trees have

8*

stood in grass ground for a number of successive years, and for some unknown reason fail to produce fruit, that if swine were confined about them for a month or so, the effect would be such on the trees as to render them productive. *Perhaps*, digging about them with a spade, and manuring, would be attended with the same result. We have our eye on several trees in that neighborhood, which bore no fruit for many years; but when the plough was applied to the soil beneath them, they brought forth good crops. The facts also furnish an unanswerable argument in favor of cultivating the soil about fruit-trees, if nothing more.

It is an excellent practice to allow swine to have the range of an orchard after the trees have come fully into bearing, as they will usually devour all the fruit that drops prematurely, and thus destroy the larvæ of noxious insects that may be in the fruit. In some instances, swine are permitted to root up the entire ground, which *may* sometimes be an advantage to bearing trees. Still, we have no faith in such a system of cultivation. If the soil were thin, resting on a compact substratum, most of the roots would be found near the surface of the ground. Hence, root-pruning by swine would work greater injury than benefit to the trees. A Western fruit-grower writes: "I have two orchards of sixty trees each; in one my swine are allowed to run from spring till the early apples ripen. A field crop of any kind is never taken from the soil. This orchard never fails of giving a full or fair crop of apples. The other is kept in grass, which is mowed annually, and no stock is allowed to run among the trees, as this can not well be done. The result is, the orchard is not worth any thing for fruit, and very little for hay."

Re-grafting old Apple-trees.—It is exceedingly difficult for any person, however capable he may be of communicating advice touching the management of orchards, to direct what shall be done with this or with that old tree, which he has never seen. The very best treatment that many old apple-trees can receive is, to cut them down at once, and remove root and branch. On the contrary, an enormous loss is often sustained by cutting down old natural trees, instead of re-grafting them, simply from a want of knowledge of the best mode of performing the work. After an orchardist has really determined to re-graft his old apple-trees, the following will be found the most approved and satisfactory manner of treating them : Before the sap begins to flow in the spring, trim all the small branches from the large limbs of the tree perfectly smooth, to a small tuft on the extreme ends, as represented by Fig. 71. This will cause the branches to throw out large quantities of suckers in all directions, most of which should be rubbed off, except in such places where you would like to place a new branch on the limb, at which points one of the most thrifty should be left. These, taking the full sap of the tree,

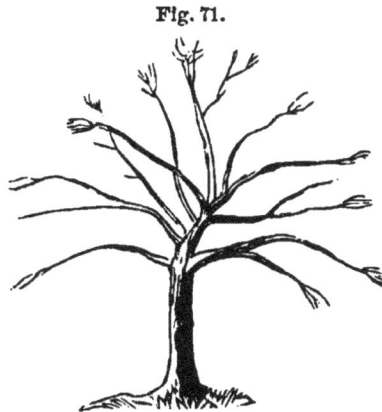

Fig. 71.

Manner of pruning an old apple-tree.

will usually, in two summers, be of sufficient size to graft, when they may be cut off and grafted, some eight or nine inches from the limb, leaving room enough for re-grafting, if the first by any means should fail. Where wounds are made as large as a cent, a coat of grafting-wax should be applied at once. After the old tree has been grafted, it will have the appearance of Fig. 72, p. 180. When the grafts are well started, perhaps the summer following, the tuft of old branches on the outer ends of the limbs should

Fig. 72.

An old apple-tree re-grafted.

all be cut off close to the grafted stock, and then carefully waxed over. In this manner one can put an entire new head on an old tree without removing any of the large branches, and can have the new branches appear in almost any place that may be desired on the old limbs. The grafts being on small stocks, soon grow over, and adhere strongly, without the danger of a rotten or defective stock; and in five years one may reasonably expect a bearing orchard, with as perfect and handsomely formed heads as fancy or skill may desire. Instead of grafting the young shoots thus thrown out in consequence of the pruning, they may be budded, if desired, the same year that the pruning is performed, provided every chance is allowed them for growth, by rubbing off a part of the supernumerary shoots. Large bearing trees may be obtained by a renewal of old ones, in much less time than by transplanting young trees. Yet, after all, one needs to exercise a vast amount of judgment as to the propriety of re-grafting old trees, as we have suggested. If they are thrifty and prolific, let them be grafted. On the contrary, if they are indifferent bearers, and are on the decline, let them be dug up at once, and a young tree started.

When one attempts to re-graft, or to renew old apple-trees by removing the old top and producing a new one, it will be found of eminent importance, after the branches have all been trimmed, to watch the development of new shoots. It is very unwise to allow all the new shoots to grow at random, and, after they have attained a large size, give the tree a severe pruning. The true way to manage such trees is to watch all the young shoots as if each one were a separate tree. After numerous buds have started, let a part be pinched off, while they are so tender that pruning can be done with the thumb-nail. Let such shoots as are to be budded be favored. A person must ascend into the tree-top frequently, to control and regulate the growth of the young shoots. A vast deal of timely care must be exercised to make the young shoots grow properly before they are budded; then, after the buds have begun to push upward and laterally, that timely care must be continued through the entire growing season.

The *modus operandi* of renewing the top of an old apple-tree by grafting is usually performed in two different ways. The first is by grafting the smaller limbs on the extremities of the branches, where they can never become any thing but mere twigs, while the body of the top must ever remain filled with the natural branches; and the fruit of the new top will appear chiefly at the extreme ends of the boughs, leaving the long, bare, and unsightly arms to annoy and exhaust the tree by its effort to throw out and support its ten thousand suckers. This is the kind of grafting usually practised by most of those itinerants who perambulate the country as professional grafters, who, being too lazy to travel, and being fairly mounted in the tree, set themselves to work, on the logical principle that, the nearer the extremities, the greater the number of branches; and the larger the number of cions they can set, the more remunerative their job will be. As a rule, such interlopers are very unscrupulous, having no regard for the success of their labor further than the development of the leaves, which is usually the standard by which the pay is regulated; neither have they any regard to the kind of fruit that is set, seldom failing of a job for the want of the variety of fruit desired. But before there is time for the fruit to prove itself, they are gone to parts unknown, to practise their deception on other localities. We would caution the public against employing this class of peregrinators as unsafe and unfit for their employment.

The other method of grafting large trees is to saw off large branches and insert the cions in these stumps, bringing the grafts nearer the body of the tree, and exempting it from the growth of the natural branches, or the pest of suckers; while it gives the cions an opportunity of attaining to some size, by removing them farther from the terminus of the capillary attraction, and bringing the weight of fruit more evenly over the tree. But this method, too, has its objections; the stocks being large, the grafts will require time to grow, so as to cover and adhere firmly and strongly to the stock. Hence, many of the young and tender grafts will be liable, while in full leaf, to be split off by the force of a strong wind; and, as they will usually come into bearing before the grafts cover the stump, they often split off with the weight of fruit; and unless there is more pains taken than is usual to keep the wounds properly waxed, the stump will begin to decay before they are grown over; and this will often prove a defect that will destroy the graft. It is a common occurrence to see a large and thrifty graft full of fruit broken square off by the wind, at the point where the cions were set in the large stock. Hence it will be found preferable, as a general rule, to trim the branches of a tree in early spring, as directed by Fig. 71, and inoculate the young shoots that are sent out on the bare limbs. By this means, a beautiful top can be formed quite as soon as by any plan of grafting. The chief aim should be to form suitable limbs at proper distances from each other, over the whole extent of the tree. It is highly important to preserve the symmetry of the tree-top. The practice of having a little bush at the extremity of a long limb on each side of the tree should be guarded against. There is a wise law of limitation to be observed in the growth and development of apple-trees. Grape-vines are sometimes trimmed twenty to thirty feet, without a bearing branch or twig. Such vines would be far more productive if the small top that produces the fruit could be placed near the root of the vine. So with apple-trees: if the little bushes at the ends of long and naked branches could be placed near the trunk of the tree, there would be much more fruit.

Every tree, when it has attained its full size, has thrown out its branches to a certain extent. But why do they stop there? Why does not the tree continue to extend its branches as long as it is vigorous and healthy? It is simply because the body and limbs are composed of small and minute pores or grains, which act by capillary attraction, in which the sap rises. Hence the tree extends to a height and distance, graded exactly in proportion to the minuteness of the pores of the wood. This is the reason why the beech, the hemlock, the spruce, and the pine each attain to a different altitude, when growing side by side in the same soil. When a branch has attained its full distance, it can not be extended by grafting beyond its own limits. Hence the importance of cutting back long and bare branches, with a view of filling up all the space in the centre of the tree-top. We frequently see a persistent effort made by Dame Nature, when apple-trees have been all trimmed out at the middle of the top, to *fill up* that vacant space by producing new shoots every season. Therefore, when all the branches but the grafts are removed, the more grafts there are to use the sap, the sooner the tree will be relieved of its pressure and disposition to form wood, and produce fruit instead. It is not essential that grafts should make just the article required for bean-poles at the end of four years, but it is important that their growth should be a medium between that of a little wood added yearly to the natural branches and the extra stimulus of a few more twigs near the body of the tree. With judicious pruning and a moderate growth, more side-branches will be sent out on the grafts, so that in a short time the outer surface of the tree will be covered with a new set of branches, completely shading the unsightly arms. When a person attempts to produce a new top on an old tree, he should know something of the art of pruning and training trees correctly. It is not enough that he possess skill to cut off a limb and set a cion in it, so that it will grow. When one has really decided to graft a tree-top, it is a common practice to graft about one-third of the top one season, another third the next season, and so on, in which case the operator commences at the top, sawing off such limbs as he desires to graft. All the branches that are to be grafted should be removed before one cion is set. It is an excellent practice, also, to saw off each limb twice. The first time it should be severed half a foot or so above the place where the cions are to be inserted. The object of this practice is to prevent injury to the stock below the grafts. The workman should always commence at the top of the tree, and work downward; then no grafts will be injured. (See *How to Graft*, p. 23.) There are so many modifying conditions and circumstances incident to producing a new tree-top, that we feel much dissatisfied with the foregoing instructions. This branch of pomology must be learned by practice rather than from books.

Fig. 73, p. 185, represents an apple-tree which is to be regrafted by installments. The branches, *a, a, a, a,* are to be grafted the first year. Those designated by *b, b,* should be grafted the second year; and those at *c, c, c,* the third year.

This illustration gives the reader merely a general idea of the details. The diagram is not a perfect model, as the centre of the tree-top has been pruned away much more than the branches should be in a real tree.

Care of the new Shoots. —The task of putting a new top on a tree is only commenced when the cions or buds have begun to grow. Some of the suckers should be allowed to

Fig. 73.

An old apple-tree-top renewed by grafting.

grow till the grafts themselves can form sufficient shade, and use all the nourishment in wood and fruit. Suckers will then cease to be troublesome. A thick coating of alburnum will be found on these bare arms, where before it hardly equalled paper in thickness. A rapid and thrifty growth will follow in consequence, thoroughly renewing the age of the tree, as well as fruit, in the least possible time. And yet, when a tree is so thrifty and luxuriant as to send out a profusion of suckers, while the grafts are growing rapidly, it would be wrong to remove them all, or to keep the growth down, lest the grafts be stimulated too greatly, and some of the sap should stagnate. Every graft on a tree requires as much timely care as a young tree. Hence the great labor of producing a new tree-top. If several cions have been set in a stock, and they all begin to grow, the best and most promising one *only* should be suffered to grow. The usual practice is to allow all to grow for a year, and then trim out the redundant branches. But such a practice is not to be commended under any circumstances.

DUCHESS OF OLDENBURG.

Fruit.—Size, medium to large; form, roundish, flattened; skin, smooth, with a light bluish bloom; color, light and deep rich red, washed, striped and splashed on a yellow ground; stem, short; cavity, acuminate; basin, deep, wide, even, regular; calyx, large, nearly closed; flesh, slightly yellowish-white, sharp, sub-acid, juicy, and, when well-ripened, pretty rich; season, September, and often keeping into October. *Tree.*—An upright, vigorous, hardy, and healthy grower, with dark-colored shoots, and broad, dark-green, coarsely-serrated leaves. A profuse bearer, apparently adapting itself to all soils and situations, and yielding a fruit of great value for marketing and for cooking purposes. It is of Russian origin.

One graft on the end of each stock will be far preferable to two or three. In many instances, there will be such an abundance of sap that sprouts will burst forth, and out-grow the grafts, unless they are pinched back. It is a very common occurrence to see grafts struggling for existence between strong and luxuriant sprouts, which choke them to such a degree that they are almost ruined. When a new tree-top is being produced, if it is desirable to develop branches which will make a well-balanced and symmetrical top, the new grafts must not be neglected when the first begin to grow. Every one should be examined frequently, and pinched, if necessary, or trained so as to grow into a beautiful portion of the tree.

Ploughing old Orchards.—Perhaps the land around the trees has already been ploughed too much. If so, plough-ing the ground again will operate like applying whip and spur to a jaded horse, after he has been urged and whipped and spurred, until he has galloped so long and so far, that the poor brute is just ready to drop from exhaustion. There are instances in which ploughing the ground of an old orchard will be the means of rendering the trees pro-ductive. Yet they are so rare, that it is safe for us to re-cord our protest against ploughing such ground, as there is a far better way to make the trees fruitful. If the land is covered with a tough sod, we would never disturb one square foot of it with a plough. The surface-soil round about such old trees is full of roots, ready to devour all the fruit-producing material that can be found in the upper stratum of the earth. The mere process of breaking it up with a plough will not promote its fertility to such an ex-tent that the trees will yield a crop of fruit. Besides this, those old trees do not need such a severe root-pruning as they will be obliged to suffer if the ground is ploughed. The far better way will be to feed the hungry roots by applying fertilizing material to the surface round about each tree, as far as the branches extend. Pomace is excel-lent when spread on the surface, as it will furnish a boun-tiful supply of apple-producing pabulum. But let not pomace be applied too bountifully near a tree, lest it de-stroy every root. A layer two inches deep will effectually destroy every thing that has life beneath it for several years; while six to ten bushels—according to the size of the tree—spread evenly and thin around each bearing tree, will greatly promote its fruitfulness. If the soil is light, sandy, or mucky, apply a dressing of marl, or clay, say one inch deep, as far as the branches extend. After this application, spread a liberal dressing of ashes and lime. There is little danger of applying too much. In late autumn, spread half a wagon-load of rich barn-yard manure around each tree. The next season, if the trees have been properly trained, pruned and scraped, there will be thrice as much fruit as if the land had been ploughed.

Straightening leaning Trees.—No fruit-tree will be so productive when the entire top and body havè been blown partially over, as it will when in an erect position. When a tree leans, sprouts will almost always start from the upper side of the body, or branches, and grow perpendicularly. By this means, the largest proportion of the sap will flow into the erect stems, while the horizontal part of the top receives only a limited supply of nourishment. When it is desirable to straighten up a leaning fruit-tree, set a post in the ground, say twenty feet from the tree, to which a chain, or rope and pulleys, may be attached, and the tree straightened up and secured with a strong wire, extending from a stake in the ground to a strap around the main part of the tree. In case a tree is large and well rooted, let the earth be excavated on the upper side and a few of the larger roots loosened, and the earth removed beneath them, so that the top of the tree may be brought up with ease to the desired position. There is no better time than early spring-time to. perform such jobs, as new roots will often start out from those that have been severed, and continue to grow till the end of the growing season. The guy-wires will always hold trees in position, until the roots will keep the top erect. If long branches grow too erèct, it is easy to wire them down to the proper position, when the wires may all be removed after one season. As a general rule, leaning fruit-trees are neglected quite too much.

When to cease Scarifying a young Orchard.—Every ambitious pomologist appears to think—as by some sort of instinct—that after a few years the land round about apple-trees must be stocked down to grass. There are instances where it would be a decided advantage to the young trees that are now producing a few bushels of apples each, to have the land stocked down. But such cases are rare. So long as the branches ripen the wood satisfactorily before cold weather, the ground should not be stocked down. On the contrary, when trees continue to grow so rampantly in late autumn that the extremities of the twigs do not ripen before the end of the growing season, it will be better to stock the ground down to grass or clover, for the purpose of checking that luxuriant growth late in the season. As has been already stated in another part of this work, let it be distinctly understood, and constantly borne in mind, that the young trees must be cultivated; the soil must be constantly stirred, and kept clean, until the orchard has fairly got under way with a thrifty growth. This is best effected by continuing the culture several years. As pomologists are often unwilling to work without an immediate return for their labor, a naked fallow among the trees will too often be neglected; but a partial crop between them will be an incentive to giving the orchard just such attention in the way of cultivation as it requires. J. J. Thomas says, touching this subject, that the length of time that this culture should be continued will depend upon the condition of the trees, and the character of the soil and surface. The orchard should have assumed the most thrifty growth before the cultivation is suspended, whether this may have required three years of culture or six. On hilly lands, with a soil disposed to wash into gulleys, we can not continue the ploughing with impunity, but must use such an alternation of crops as will obviate the necessity for constant open culture. This may be arranged by a rotation of clover with corn or potatoes, which will be a valuable alternation, since this legume is itself almost a cultivator of the soil, rendering it loose and mellow, while, at the same time, the surface is clothed, and the soil is bound together by its roots. Moreover, the red clover plant attracts much of its sustenance from the atmosphere through its abundant foliage, and the radicals sink deep into the subsoil in search of nutriment. It will always be found a safe rule to cultivate young orchards until the trees are ten to fourteen feet high, and of a corresponding proportion as to breadth of top; after which, stock the land down, and mulch the surface around the trees, if they require it, instead of scarifying it.

Neglected Apple-trees.—The country is replete with such disfiguring blotches of the homestead or orchard. A neglected apple-tree appeals more strongly to the eye of the pomologist after it has shed its leaves; and in too many instances the tree is hastily cut away, leaving but its stump as an unpleasant reminder of the loss of a valuable old friend, which might have been saved by a little timely care. Where the early training of a tree has been neglected, there are often several leaders forming the top. In this case, an acute bifurcation is to be carefully prevented by checking and shortening the weaker side. If we neglect this, there is danger of the tree splitting from top to bottom when loaded with its first heavy crop.

Every fruit-tree should grow a little every season, as the growth of new wood *every year* is Nature's provision for the healing and repair of wounds, for the increase of size in trees, and for productiveness. If a fruit-tree do not grow at all, it can not be expected to produce a crop of fruit.

Hence, when a tree has been neglected so long that the entire top has become a dense hedge of brush, with the branches all intertwining so closely that a person can scarcely climb into the top, the first thing to be done is to commence the work of thinning out the supernumerary branches. Perhaps Dame Nature has already begun to prune in her own slow manner, as she prunes forest-trees, by withholding the annual flow of sap from certain branches until they die and drop to the ground. A good pruning-saw will be found the most convenient instrument for this purpose. Do not make a clean sweep of every branch, but thin out gradually, here and there, cutting away the poorer boughs first. Let the aim be to thin out the crooked and weak twigs and branches first. Where two limbs cross each other, so that one rides on the other and interferes with its growth, let the poorest one be cut away. Let the centre of the top be thinned out to such an extent that a person can climb around among the boughs. If the extremities of the limbs are very bushy, take the pruning-shears (Fig. 53), and thin out the old twigs, so that new spray may be produced, which will bear fruit. Let the branches be cut off smoothly, as directed on page 128, Fig. 57, at *a*, after which take the kettle of liquid grafting-wax (see p. 55) into the tree-top, and smear all the wounds. This treatment will give a neglected tree such a start, that, if the soil is rich, a crop of fine fruit may be expected in two years after the pruning.

Renewing Stunted Trees.—It frequently occurs that apple-trees—and the same is true of other fruit-trees—after being removed from the place where they sprang from the seed, or from the nursery-row, fail to grow at all satisfactorily, and, in many instances, they barely live from year to year. The difficulty in such cases is, the trees were too old to be taken up with removing a ball of earth without the roots; or the tap-root and a large proportion of the laterals were severed, and left in the ground when the tree was re-moved; or the tree was taken from a rich and deep soil, and transplanted in very uncongenial and poor ground, where there was perhaps no surface mould at all, and where the entire ground is hard and almost impenetrable by roots of any kind. Any one of these conditions or circumstances will be sufficient to give a tree an almost fatal stunt. In many instances, the sub-soil is so thoroughly saturated with water, that the roots can not spread downward.

Fig. 73½.

A SLUICE IN THE ORCHARD.

When stock of any kind are permitted to feed in an orchard where all the water is collected in under-drains, a sluice, like Fig. 73½, will be found a source of great convenience in providing clean water for them to drink. A plank box, six to ten feet long, without bottom or top, is sunk, as represented, so that the water, flow-ing through a drain, may enter the box at one end, and run out through a hole at the other end into the ditch, and pass off. Sheep and swine can walk down on either side of the sluice, thrust their heads into the large holes in the side planks, and drink without fouling the water. Horned cattle and horses can also drink at the same watering-place. The ground should be excavated on both sides, and paved with small stones, so that the feet of animals will be as low as the water in the sluice.

Management of young bearing Trees.—As soon as a well-trained apple-tree begins to bear, the only pruning required is the annual reduction of the leader, in order to thicken the trunk at the base, removing any rival shoot that may threaten to compete with it, and taking out a little useless spray from the centre of the tree, if the branches are too thick and are in each other's way. One great object being always, in pruning, to promote hardiness and fertility, these two points must be always kept in view. As soon as an apple-tree is able to bear, permit a moderate production of fruit, but do not allow a heavy crop. It is better to have a few fine, well-formed apples at first, near the body of the tree, than a lot of miserable half-grown specimens dangling in the wind at the end of the limbs. Wherever a tree has been so neglected that it has two or three leaders which can not be cut down to one, without wounding and nearly girdling the limb selected to remain, it is better to retain them all, and prune the whole into a compact oval or distaff form; and, therefore, the limbs spreading at a right angle should be carefully reduced, not immediately at the trunk, but at a convenient distance, some few inches from it; for it is better to make an elbow—which, in pruning, is generally to be avoided—than to leave a branch too long whose length will destroy the regular oval shape of the tree.

In many instances, there is such a desire to have young trees produce a bountiful crop of fruit, that a young orchard is kept in an unthrifty condition for many years, from no other cause than overbearing while the trees were growing rapidly. Our own practice has ever been to clip off, in many instances, seven-eighths of all the young fruit, as the growth of the branches was of more importance and value than the small quantity of fruit that would be pro duced at the expense of the future hardiness and vigor of the bearing shoots. It will be found more injurious to young trees to bear a bountiful crop of fruit than for young animals to produce young before they have attained a proper age for propagating their species.

CHAPTER VII.

THE FAILURE OF ORCHARDS, AND THE PRACTICABLE REMEDY.

The price of our apples, of peaches, and cherries,
The price of fine currants, of pears, plums, and berries,
Is measured by combats with foes in a tilt,
With war to the knife, and the knife to the hilt.—EDWARDS.

THE failure of apple-orchards is a theme of common con-
versation in all localities where apples are cultivated. In
the majority of instances, pomologists *assume* that apple-
trees and pear-trees fail to produce such crops of fruit as
were once raised in certain localities, because the varieties
are ˙ running out. They feel confident that they have
guessed correctly on this point. But the assumption is a
great delusion; and we can convince any intelligent pomologist that the conclusion is erroneous. If the same quality
of soil can be secured, and the same surroundings as to
protection by forests, and if the cions from the topmost
boughs of old trees that have once borne bountiful crops
of fine fruit, but have now failed, could be set in young
stocks as hardy as the stocks were which were employed
sixty or eighty years ago, we should see trees loaded with
just as fine fruit as the old trees ever produced. The truth
is, there is a limit to the productiveness of all kinds of fruit-
trees. Old animals cease to bear. And why should not
trees grow old and barren? Forest-trees reach a limit in
their growth, and decay. And can it be expected that apple-trees will continue to bear after they have passed the
maximum limit of life? Durham cattle die; but the *breed*

—*the variety*—does not deteriorate. If old trees cease to bear, and die, the *variety* of fruit does not fail, if cions from the branches be set in young stocks. If a piece of ground be cleared of the forest, by burning every thing to ashes on the land, as farms used to be cleared, and heavy bodies of growing timber be left standing, to protect the young orchard, there will be no complaint of failure of fine fruit. The trouble is in the culture and treatment of the growing trees. The young orchards that ought now to be yielding bountiful crops of fine fruit have been produced and cultivated in the most perfunctory manner. And then, because it is impossible for trees to bear under such circumstances, pomologists have *assumed* that there is "something in the air—some pernicious east wind that blasts the embryotic fruit, or the varieties are running out!"

1. Now then, let us sift the evidence, point by point, and take an honest view of the subject, by comparing the past with the present. Four-score years ago, the stocks into which cions were set were produced from more hardy varieties than the stocks of the present day. *Now*, nurserymen plant the seed of *any* kind of fruit. It can not be denied that the stock exerts a wonderful influence on the productiveness of the bearing tree.

2. Four-score years ago, fruit-trees were set in a *virgin soil* which, in most instances, had been top-dressed bountifully with unleached ashes, which are an almost indispensable requisite in the production of fine fruit. But *now* inferior trees are planted in an inferior soil, *without* wood ashes and other fertilizing materials, which growing fruit-trees *must* have before they can produce a bountiful crop.

3. Four-score years ago, almost every orchard was shielded by a belt of forest-trees. But *now*, cold and fierce winds sweep over the country for a long distance, raking young orchards in a fearful manner.

4. A large proportion of the orchards all over New England, and many other States, have been almost ruined by stupid management. As the soil had become very much impoverished, in consequence of having yielded a long succession of bountiful crops of fruit, and the trees began to show the results of injudicious management, some stupid wood-chopper would be directed to prune the trees; which would be performed by cutting off about half of the entire top, and in numerous instances *more* than half; and large limbs, six to eight inches in diameter, have been chopped off, making such broad wounds that they could never heal over. Hence, the wood began to decay, so that thousands of valuable apple-trees, that once yielded full crops of fruit, decayed to the heart, often becoming hollow. Such management and treatment of fruit-trees is surely sufficient to effectually annihilate the most hardy and prolific variety of pears or apples that has ever been cultivated.

The diagrams of two murdered trees (Fig. 74), which were originally engraved for the " Annual Register," will furnish a fair idea of the system of management adopted by untold numbers of farmers. The trees have been ruined by some *tree-murderer*, who has cut and carved, right and left, leaving long and uneven stubs, which, instead of healing over, have decayed. Besides this, the land has been cropped until the soil has no more fertility left in it, either for producing apples or any

Fig. 74.

Apple-trees badly managed.

other crop. Large sprouts have been allowed to grow from the collar, and the trees have received such ruinous treatment that they are nearly worthless. Let the reader compare these ruined trees with the fine-appearing illustration on p. 117. An axe is no more suitable for trimming apple-trees than such an edge-tool is for cutting one's finger-nails or his hair.

5. Four-score years ago, noxious insects, grubs, beetles, curculios, borers, caterpillars, worms, and aphides, seldom injured bearing fruit-trees more than enough to thin out the superabundant crops. But *now*, if a man plants an orchard of the very best and hardiest varieties of apples, he must watch for grubs in the roots, for borers in the bodies of the trees, dispute his title to the fruit with innumerable caterpillars, with army-worms, canker-worms, and many other worms, with untold numbers of the codling-moth, the irrepressible curculio, in a hand-to-hand contest, and with uncounted numbers of aphides, which are ready to suck the last drop of sap from every growing tree.

From the foregoing suggestions, every intelligent pomologist will perceive what difficult influences and formidable enemies of both the apple-trees and the fruit must be encountered when an effort is made to produce a crop of apples.

We well recollect, when a small lad, that the father of the writer, and most people in that vicinity, had their orchards cut to pieces in the most stupid and wanton manner, by men who knew no more about training and pruning fruit-trees than the verdant son of Erin who *cut down* a young orchard when he was directed to prune the trees. We have seen many large and prolific apple-trees, after the bodies had literally decayed to the heart, in consequence of the large wounds, standing on two prongs before the trees died. None of those orchards that were hacked so unwarrantably, ever recovered from the fatal injury of such stupid pruning. Is it any wonder that fruit-trees fail to bear, when treated thus? Is it not rather a wonder that varieties maintain their identity as well as they have done?

Worthless Trees.—In addition to the foregoing causes of failure, there are other causes of serious failure in the production of apples, which a beginner must understand before he can calculate on success. With nurserymen who have trees to sell, and who have no further anxiety than simply to produce trees which will command a high price, a tree that will never produce a satisfactory crop of fruit is just as good as one that will yield annually a bountiful crop. There are untold numbers of worthless apple-trees sold every season, which will never yield one-fourth the amount of a fair crop; and yet, to all external appearances, they were, when transplanted, smooth, thrifty, and superior trees. If a tree has not been produced properly, it can never be expected to yield a satisfactory crop of fruit; and, when apple-trees that *have* been produced as such trees should be grown, are not managed correctly, they will not* yield a supply of fruit. The nursery business has been engaged in by such a large number of persons, who are utterly ignorant of the fundamental requirements in order to produce bearing trees, that our country is overstocked with worthless fruit-trees, which *appear* all right; but which, with superior management, will never yield a satisfactory crop of fruit. Every intelligent, *practical* pomologist, if he has exercised himself to observe what influences affect growing fruit-trees, both favorably and injuriously, knows that an apple-tree which has been produced by setting a cion in a piece of the root of a young tree that has sprung from the half-ripe seed of a miserable apple, can never be made to yield satisfactory crops of fruit.

Woodpeckers.—There are several kinds of woodpeckers of great value to pomologists, namely, the Red-headed Woodpecker, the Hairy Woodpecker, and the Golden-winged Woodpecker. Fig. 75 represents the parent bird feeding a grub to a full-grown fledgeling. At the lower part of the picture, a young woodpecker, almost large enough to try his pinions, is represented as looking out on the world from his nest in an old tree. Although woodpeckers will eat good fruit, they should not be destroyed, as they perform an excellent service in exterminating noxious insects while the fruit is growing. We have never known them to do any other

Fig. 75.

RED-HEADED WOODPECKERS.

damage than to eat cherries and mulberries. They are so shy, that it is not difficult to frighten them away from fruit-trees by means of a few men of straw, suspended from the end of a long pole, so that the wind will keep the images in motion. The length of this bird is nine and a half to ten inches; alar extent, seventeen inches; bill, light blue; legs, bluish-green; iris, dark-hazel; head, neck, and throat, crimson; back, wings, and tail, black, with bluish reflections; secondaries, rump, lower part of the back, and under parts of the body, white. Female, less brightly colored; young, head and neck dull gray, varied with blackish.

The Red-headed Woodpecker is well known throughout the United States. In spring and summer, it is seen in almost every region. Their return northward in Pennsylvania is about the 1st of May. About the middle of that month, they prepare their nests in the large limbs of dead trees, adding no materials to the cavity which they smooth out for the purpose. Sometimes several perforations are found in the same tree; but living trees are seldom occupied by them. The same tree is employed for years in succession by a pair of these birds. The eggs, usually six in number, are white, marked at the largest end with reddish spots, in which last particular they differ from all others of the genus. The first brood appear about the 20th of June. Both the eggs and young of this, as of many other birds, often fall a prey to the common black-snake. These birds are exceedingly agile, securing with ease the beetles seen from their perch. When the fruit is all gone, their facility in detecting insects under the bark of trees is remarkable. Alighting upon the trunk, one of them will stand motionless for a few moments; then he will strike the tree with his bill, and seem to be listening to hear the sound of the borer's auger in the bark, or an inch or more in the wood of the apple-tree. Woodpeckers seem to have been made with an especial reference to the work of destroying borers in all kinds of trees. After a borer has worked his way two inches into an apple-tree, unless he plugs his passage tightly after him, the woodpecker will thrust in his long, harpoon-formed tongue, as represented by Fig. 76, and haul the fat borer out and devour him.

Fig. 76.

The Hairy Woodpecker, which is often known as the "Sap-sucker," is frequently killed on account of the erroneous notion, cherished by some persons, that this bird pecks holes through the bark of fruit-

A wood-pecker's head with the barbed tongue extended.

trees, to feed on the *cambium* (see Glossary), when the faithful little bird is taking out grubs. It is a common occurrence to see hundreds of holes like the depression of a large gimlet made in the bark of trees, at the bottom of which the bird found a grub. Had those depredators been unmolested, they would have made a complete honey-comb of the body of the tree in two or three years. Through profound ignorance, many persons often destroy, in a relentless manner, some of their most faithful benefactors.

The Golden-winged Woodpecker—called also the Yellow Hammer, High Holder, Yucker, and Flicker, in other parts of the Union, being seldom known by the name of Golden-winged, employed by naturalists—may be said to be one of the least destructive of the birds regarded as injurious to agriculture, while it lives to a great extent on insects and borers that infest fruit and other trees. This bird is especially recognized by its *flicker, flicker, flicker, flicker*, which, at a little distance, is like the sound made by a mower as he whets his scythe. At all times animated and happy, these birds are peculiarly so at the love-making season of early spring, when their voices may be heard in the utterance of joyous sounds, and when the coy female is pursued by several males until she has indicated her preference, which produces no strife, as the rejected lovers at once fly off elsewhere to woo. The song of the male, at this season, is not unlike a jovial laugh, nor by any means unmusical. As soon as mated, each pair immediately proceed to excavate the trunk of a tree, and to fashion a place for themselves and their young. The hole is at first made horizontal, and then downward about six or eight inches. They caress each other on the branches, climb about and around the tree with apparent delight, rattle with their bills against the tops of the dead branches, chase away the red-heads, and feed abundantly upon ants, beetles, and larvæ. Before two weeks have passed, from four to six semi-transparent eggs are laid. Two broods are thus produced in each season. The movements of one of them upon the side of a tree or upon the ground are very quick, though it only alights upon the earth to pick up a beetle, caterpillar, or other insect, or perhaps a grain of corn. The young of this species frequently have the whole upper part of the head tinged with red, which, at the approach of winter, disappears, when merely a circular line of that color is to be observed on the hind part, becoming of a rich vermilion tint. Some writers have denounced this bird as unfit to live. But we have seen no evidence that it is not a useful bird.

Fig. 77.

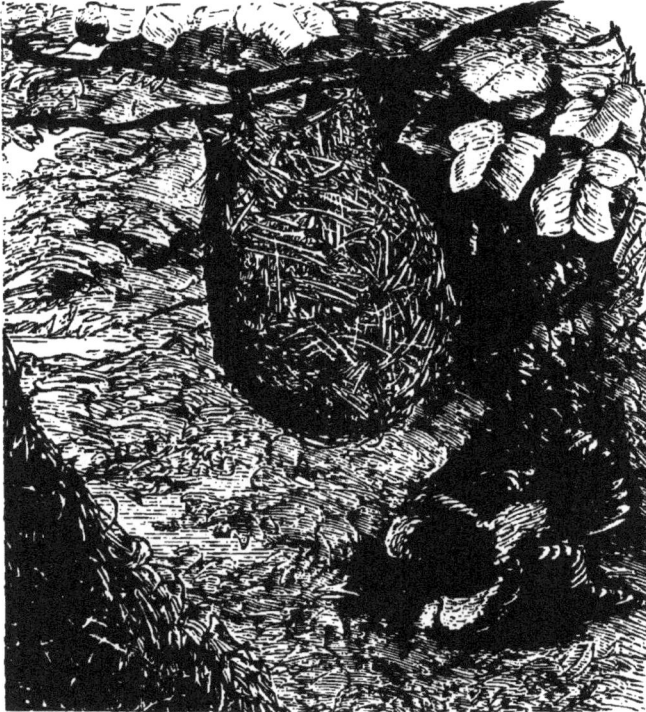

NEST OF A BALTIMORE ORIOLE.

The Orchard Oriole is one of the most valuable birds that pomologists can encourage to dwell among their fruit-trees, as they subsist almost exclusively on noxious insects. The Baltimore Oriole is another bird of great value in apple-orchards. It is exceedingly difficult for crows or hawks to take the eggs or the young of these birds out of their nests, as every nest is usually suspended beneath swaying branches, as represented by Fig. 77. A pair of Baltimore orioles were accustomed to build their nest every year on a favorite pear-tree near the writer's dwelling, in Central New York, several years ago; and it was an admirable sight to observe the wonderful skill displayed by those birds in building their nest. The first step was to procure a piece of wrapping-twine, one end of which was wrapped three times around a horizontal branch just beyond the fork of a branch, and the end was tucked beneath the main strand, so that the greater the stress or weight on the string, the tighter it would clasp the branch. The other end was then wrapped about the opposite branch of the fork, allowing the middle to hang

9*

down in the form of a basket or an oblong pouch, Fig. 77. Several such strings
were attached to the branches with wonderful precision. Strong spears of grass
were also employed and woven in among the pieces of twine, as a basket-maker
weaves or braids a basket. All this is performed with the bills. We could pen
pages of interesting matter touching the natural history of the orioles; but we
have only space to implore farmers, and every one, not to "destroy the goose
that lays the golden egg." Spare the orioles. Encourage them to abide in the
fruit orchard. Do not allow ill-natured boys to stone them, nor cruel sportsmen
to shoot them. They will destroy an immense number of noxious insects. Let
children be taught in early life that the orioles are their best friends, to whose
efforts many a fine crop of fruit is justly due.

Fig. 78.

GOLDEN ORIOLES AND NEST.

The Golden Oriole, illustrated by Fig. 78, is one of the useful little birds that
flourishes for a few months during the former part of the growing season near
human habitations. Birds of this variety seldom feed on fruit. They carry large
numbers of worms and bugs to their young ones; and their chief food, at all
times, consists of noxious insects.

Apple-trees starved out.—That inquirers may perceive what views other writers entertain touching the cause of failure in orchards, we herewith quote from the "Canada Farmer," the editor of which says: It is no uncommon thing to meet with single trees, or even whole orchards, once productive, that have become barren, either yielding no fruit whatever, or a few scrubby specimens—mere apologies for the products formerly given. In such cases you will perhaps hear it said that the trees have failed, or become exhausted; whereas the truth is, that the *soil* is exhausted. Such phenomena are viewed as among the mysteries of the vegetable kingdom; but they are among the simplest and most easily explained facts to be met with out of doors. It would be difficult to find any thing more irrational than the course pursued by many in regard to fruit-growing. A young orchard is planted out, and forthwith sown to a grain crop, in which the trees stand during the summer months, like storks in a rush-pond, their heads just peering over the nodding grain. Year after year a similar course is pursued. The land is expected to bear as much of some sort of crop as though an orchard had never been thought of. After much hard struggling, half the trees, perhaps two-thirds, are found to have survived, and they begin to bear a little fruit. At last, by a stretch of leniency, the orchard is seeded down; and, after one or two mowings, converted into hard-run pasture. Who ever thinks of manuring an orchard? or what fool would dream of giving up the land to the trees, and manuring it well, and cultivating it thoroughly? Yet if, after the worst possible usage, the trees do not bear plenty of choice fruit, either the nurseryman is blamed, or the climate is cursed, or, forsooth, the trees are exhausted! The theory of rotation of crops, in general farm practice, is based on the fact that constantly growing the same produce exhausts particular

clements in the soil; and this fact, no doubt, often accounts
for the barrenness of fruit-trees and the failure of orchards.
The soil is robbed of its nutritive properties year after year
—no new supply is furnished—and out of nothing, nothing
can come. The practical lesson is obvious: we must feed
our fruit trees if we expect them to feed us.

Itinerant Grafters.—Taken as a class, they are no more
reliable than Jew peddlers who are selling stolen goods.
They have wrought irreparable mischief in many a thrifty
orchard, by cutting off such a large proportion of the
branches, that almost every tree received an injurious—almost fatal—set-back, by having too much of the top removed at one time. In almost every part of the country,
people have been outrageously deceived by itinerant grafters. They traverse the country, and take orders to do
grafting at so much apiece for all that live, or at so much
for each cion set, ready pay. When the season of grafting
comes, a few workmen come along with a wagon-load of
cions, containing, professedly, every variety that could pos-
sibly be called for, all procured from the most responsible
source; and as a proof of this, a catalogue of some well-
known nurseryman is exhibited, and, it may be, a forged
bill or invoice, while the cions were, in many instances, cut
from some of the orchards they had been grafting in. Thousands of orchards have been ruined in this way. We have
in mind an orchard which has been grafted by one of these
rogues, and, instead of having some three or four select
sorts, he had a collection of vile rubbish, mostly natural
fruit; and, in some cases, three or four different sorts were
grafted on a tree. We might cite cases of this sort which
have come to our knowledge, enough to fill a dozen pages.
In some parts of the country, people are more cautious and
careful than formerly, so that few men now are willing to
trust unknown, irresponsible persons with the important

duty of grafting their fruit-trees. Not so, however, in some parts of the West and South, where, we are informed, the speculation is in full blast. It is but just to say, in this connection, that there are honest men engaged in this business of grafting—men in all respects worthy of confidence —and the service they render to fruit-culture is very great. What we have said will be no detriment to them; for they have characters to sustain them and to inspire confidence.

The remedy for this source of injury to an orchard is— when trees are to be grafted—to procure cions from thrifty, bearing trees, during the winter, and let the grafting be performed by some intelligent person who understands something of the science of pruning fruit-trees.

Want of proper Pruning constitutes a prominent cause of failure in the productiveness of many apple-trees. When some kinds of apple-trees are permitted to grow *ad libitum*, as if branch-extension were the great desideratum, and expend all their forces in the production of wood-growth, they can produce little or no fruit. Indeed, it is not possible for any tree to perfect a fruit-germ, and not again in some way disorganize it, unless the wood-growth shall cease in time for the leaves to elaborate a sufficient quantity of plant-food to grow both leaf and fruit the following year, or until a part of the leaves shall attain to nearly or quite their full size. That this is so, will be apparent when we consider that the leaves which first appear in the spring were formed in the buds the previous year, perfect in all their parts, and in the embryo state contained each individual cell found in them when fully grown. As growth begins in the spring, the small cells which were formed in the previous year begin to expand. Each individual cell thus enlarges, until all the cells of which these leaves are composed have attained full size. Here, for example, is a tree in possession of a full-grown leaf. This leaf did not form itself, but was formed by the tree in the preceding year. To produce and sustain this cellular enlargement, there has been stored, the previous year, a large share of nutriment in the buds and in other parts of the trees. This nutriment must not only be sufficient to feed the embryo leaves, but also be sufficient to produce the small warty excrescences—the rootlets and spongioles. These new leaves and spongioles are the tree's laboratory. The first growth of leaves and spongioles was made, with the exception of moisture, wholly out of the materials that were stored by the tree during the growth of the previous year. When these vegetable stores are in sufficient supply to do this, and nourish the fruit-germs also, then we shall hear little about imperfect fertilization. On the other hand, had the food been consumed the previous year, by ripening the over-crop of fruit, or by making a very succulent growth, then the tree would not have a sufficient amount of plant-food to perform its three-fold office in the production of leaves, roots with their spongioles, and fruit. In this condition, a part of the leaf, and a larger part of all the fruit-buds, yield up their nourishment, which goes to the production of root and leaf growth. The tree, therefore, is barren of fruit for the summer, its whole growth being required to recuperate its own vigor. Such trees bloom freely, and then cast their blossoms. When this occurs, uninformed persons often attribute it to want of fertilization, or suppose that the rain must have washed away the pollen. The true remedy is, adopt the means recommended, p. 167, for the purpose of promoting fruitfulness.

Want of Protection in Winter.—We are aware that round about New York and Boston, and other localities, where orchards rest in the very bosom of the sea, the pomological authorities scout at the idea of protection. In many localities in Southern latitudes, fruit-trees do not require any protection from winter winds and cold storms. Hence we do not advocate universal protection. But, at a distance from the ocean, in most localities, apple-orchards have suffered extremely—often fatally—for want of the protection which a belt of timber will afford.

Lewis Ellsworth, who is acknowledged to be reliable authority at the West, writes that the loss in fruit-trees in Illinois, within the last three years, is *millions of dollars*, which is attributed to the cold winters and dry summers. But he asserts that, to a great extent, this result has arisen from their standing *unprotected* in a soil underlaid with a retentive clayey-loam subsoil, which characterizes most of the prairie lands. He has adopted the practice of ridging his land by repeated ploughings, commencing at the same ridges, and ending at the same dead furrows. Where nursery-trees were formerly thrown out by freezing, since ridging, they stand throughout the winter without injury, and make a better growth in summer. He recommends the ridging system for all orchards, each row of trees being placed on the centre of the ridge.

We have no doubt that the thorough draining of such ground would lessen the effects of severe winters on fruit-trees in other regions as well as at the West.

Belts of Timber around Fruit-orchards. — We will suppose, for example, that an orchard is to be planted on a broad prairie, where the cold winds of a North-western winter can rake the surface of the earth for twenty to thirty miles ; and suppose, farther, that a few neighbors are in possession each of fifty acres of land. Let them all co-operate in establishing belts of forest-trees as effectual wind-breaks, around every fifty-acre plot. By working in concert, as joint owners of land build line fences, heavy belts can be established in a few years.

The first step should be to break up a strip of ground

not less than four rods wide, early in autumn.   Let the
ground be ploughed deep.   Then scarify the surface with
cultivators and harrows, until the whole strip is reduced to
a fine tilth.   Now stick four rows of stakes about one rod
apart, lengthways of the plot, where rows of trees are to
grow.   The two outer rows will be about eight feet from
the border.   Hoed crops may be cultivated on such ground
for a few years.   Or the surface may be worked occasionally with a horse-hoe, to subdue the grass and weeds, and
to keep the soil moist.   Crops of cereal grain should never
be raised where such trees are growing, as the aim should
be to promote the development of every tree, by keeping
the soil moist and free from vegetation during the growing
season.   If grass and grain are not allowed to grow around
such trees, every one will shoot upward, in good soil, from
two to four feet annually.   In ten years, by proper cultivation, a dense belt may be formed thirty feet in height, entirely around a fruit-orchard.   But let grain and grass be
cultivated among the trees, and the tops will not stretch
upward more than one foot in each year.

What Trees to cultivate.—The aim should be to select
such varieties as can be propagated easily, which will
grow rapidly, mature their branches thoroughly before cold
weather, be hardy during the winter, and make valuable
timber.   Among such varieties may be mentioned the butternut (*Juglans cinerea*), the black walnut (*Juglans nigra*),
the yellow locust (*Robinia pseudicacia*) ; in some localities
the chestnut (*Castanea vesca*) will flourish satisfactorily.
Some Western pomologists recommend for this purpose,
Cotton-wood, Tree Willow, and Silver Leaf Maple, which
will in a few years make an effective screen for an orchard
and farm crops, all of which will grow rapidly, and most of
which will yield durable timber.   There are also other trees
which would be, perhaps, preferable to these in certain lo-

calities. But, whatever the varieties may be, the trees that will grow most rapidly and the taller should always occupy the middle of the belt. If slow-growing varieties be planted in the middle rows, and rapid growers on the outside, the trees in the middle of the belt will soon be overgrown, and their natural development choked down.

Just before cold weather, mark out the ground in the quincunx order, and plant the nuts or seeds about eight feet apart, except the two outside rows, which should consist of hardy evergreens that can be obtained most conveniently from nurseries or the forest. Hemlock, cedar, and pine will grow rapidly, if every one has a good tap-root. Hence it will be important to select very small trees, which have a tap-root. The same is true with all nut-bearing trees. It will be found far more satisfactory to plant the nuts where the trees are to grow, so that the tap-root may not be injured, than to attempt to transplant trees from nurseries. A protecting belt can be formed much sooner, and the trees will all be of a more uniform height, by preparing the ground as suggested, and planting the nuts and seeds in autumn, than by taking trees from the nursery. Pine-trees will seldom flourish satisfactorily without the aid of a tap-root to bring up moisture in hot and dry weather.

Necessity of Protection.—Let the reader understand that we are not advocates of *universal* protection. It is only advocated in localities where deep snows fall, where protracted cold weather and piercing winds prevail, and where the surface of the earth round about trees freezes up so tight and to such a depth as to check all the circulation of the sap in the roots. Where trees and the ground are thawed out the next day after freezing up tight, no timber-belts will be needed. We should remember that apple-trees are things of life—living and breathing existences, as much as many other specimens of animate nature, which

RED ASTRAKHAN.

Synonyms.—Deterding's Early, Astrakhan Rouge, Rother Astrakhan, Abe Lincoln, and Vermillon d'Ete.   This beautiful apple has come to us, through England, from Russia. Its introduction into this country is quite recent; but it is now seen in our markets, and sells as well as any other variety of the same season.   It generally bears well, and is in all respects worthy of a place in every collection.   Fruit is rather above the middle size, very smooth and fair, roundish, a little narrowed towards the eye. Skin almost entirely covered with deep crimson, with sometimes a greenish-yellow in the shade, and occasionally a little russet near the stalk, and covered with a pale white bloom. Stalk deeply inserted. Calyx set in a slight basin, which is sometimes a little irregular. Flesh quite white, crisp, moderately juicy, with an agreeable, rich, acid flavor. Ripens from the last of July to the first of September, according to locality.

are endowed with powers of locomotion.   A vital fluid circulates through every radicle and hair-like rootlet, through the stem, every branch and minute twig, thus maintaining the healthful existence of every bud within the folds of which is enveloped the future fruit, or nothing but a leaf. Countless numbers of mouths are ever open to catch the pearly rain-drops and the minute tidbits in the well-prepared soil, which are conveyed to the buds to expand the blossoms and develop the luscious fruit.   The genial breezes of the spring, the glowing sunlight, the refreshing showers of summer, the pelting storms and intense cold of winter, all exert more or less influence on the vitality of a fruit-tree. There is more or less circulation of the vital fluid through the roots, stem, and branches, at all seasons of the year. Although the circulation may be feeble during the reign of stern winter, when the whole vegetable kingdom is bound in icy fetters, still there is a living, breathing, active vitality in every branch, twig, and bud.

Any one who has access to a growing tree may satisfy himself that this suggestion is not based on some groundless theory. By severing a live limb from the parent stem, covering the end with wax, and securing the amputated branch in the tree-top, where it will be exposed to the same influences of alternate heat and cold, it will be seen, after a few weeks have passed, that the severed branch is drying up. This fact assures us that there is a constant transpiration through the bark and buds. Consequently, if the branches and buds do not receive a supply from the roots, the last drops of moisture will be taken out of the green bark by the piercing and drying winter winds. We all understand how rapidly the cold and driving winds of winter will suck out and convey away the last atom of moisture in a wet board or a wet garment. When "Old Boreas" has drawn his big fiddle-strings up to concert pitch, so that swaying forests keep time with his music, which makes the vast plains roar, and hill and valley echo with the sound of a tempest, the searching, driving, and drying wind will have the moisture of the trees, if—to use the emphatic phrase of a boorish rough—the last drop "must be taken out of the hide." Hence, if the ground be frozen up solid, as far down as the roots extend, every mouth must be stopped; and, consequently, the supply of moisture for the buds and twigs will be cut off. Then, if cold and drying winds prevail for many successive days and weeks, who does not perceive that the very life-sap of a growing tree must yield to the imperative demands of the furious wind? This perpetual draught of the drying wind on the scanty supply of sap and moisture, which is sucked out through every pore of the buds and bark, exerts such a deleterious influence on the productiveness of the fruit-tree, that the embryotic fruit is destroyed, and the vitality of the tender branches so seriously impaired that the trees can scarcely survive during the growing season. These suggestions furnish a philosophical reason why it is found so exceedingly difficult, in the West and North-west, to obtain fruit-trees that will endure the rigors of our Northern winters. Every intelligent pomologist will readily understand and appreciate the conditions, circumstances, and influences to be encountered if one attempts the cultivation of a fruit-orchard.

The question will then arise, Is there any practical remedy? And is it available? Yes, perfectly so. If a plot of ground be cleared in the centre of a forest, and an orchard planted, the fury of the drying winds will be spent on the breast-work of forest-trees, and the draught for moisture will be less severe on the buds and branches. The ground, also, will not be frozen to so great a depth. And, as a consequence, the circulation of the vital fluid will not be so completely obstructed. Hence the embryotic fruit will not have the last drop of moisture sapped from the dormant buds. Therefore the vitality of a fruit-tree will suffer but little injury from protracted periods of cold and drying weather. All these conditions and circumstances are controlled and regulated by natural laws, which are as reliable as the vicissitudes of the seasons. And it is the business of pomologists to study the operations of the laws which affect vegetation, and to make an application of the knowledge gained in one thing to the successful accomplishment of another purpose.

Now, then, if pomologists will observe all the fundamental requirements in the production of fruit-trees, by selecting the best kernels for bearing trees, and cultivate and train in strict accordance with the laws of vegetable physiology, then encircle their young orchards with a complete wind-breaker of trees, and cover the ground round about the trees with straw or sawdust, to prevent the frost from closing the tens of thousands of minute feeders, which must supply the branches and buds with moisture, even in cold weather, such an occurrence as a failure of fruit, in consequence of the severity of a long and cold winter, would be rarely met with, save in the records of the past. We have the cheering promise of seed-time and harvest while mortal life is prolonged. Seed-time has never failed. If the fruit-harvest is cut off, we must look for the cause of such a failure in the system of management adopted in the production and care of the trees.

Many of the wise pomologists of America are beginning to wake up to a proper appreciation of the serious mistake of their fathers in denuding their farms of the last acre of forest-trees; and many are beginning to repair this great injury done by the stupid managers, by establishing belts of deciduous and evergreen trees entirely around their farms, to protect not only their fruit-trees, vines, and bushes, from the scathing influence of cold winds, but with a view of shielding the winter crops of the field from the same influences which have nearly cut off "the staff of life" in those localities where bountiful crops of golden grain were harvested, so long as the tender wheat-plants were protected from the drying and withering influences of cold winds by the surrounding forests.

Management of Girdled Trees.—The true way in the

Fig. 79.

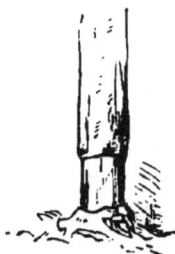

A stem of a tree that has been girdled by mice.

management of orchards is not to have any trees girdled. But sometimes mice, sheep, pestiferous goats, or rabbits, will gain access to fruit-trees, and girdle the bodies in a few minutes. A friend of mine in Central New York, who has several thousand fruit-trees, had over one thousand young standard trees completely girdled by mice during the latter part of March, while a deep snow was lying on the ground. Previous to that time the mice had never injured a tree. Some of the trees, which were from one to two inches in diameter, were girdled over a space of six inches above the surface of the ground. Such casualties are of frequent occurrence. His trees were all saved but sixty; and those

Fig. 80.

A stem of a tree "bridged" after it has been gnawed by mice.

would have survived if they had not been submitted to a bungler. The accompanying illustration (Fig. 81) will show how the girdled portions were bridged with pieces of living branches taken from the tree-tops.

Fig. 81.

A young tree four years after the girdle was bridged.

The main idea is to insert four or more pieces of live branches in the stem of the tree, with one end above the girdle, and the other end below it, so as to maintain the connection between roots and stem, which has been cut off by the removal of the bark. In many parts of the country field-mice are so numerous that, unless fruit-trees are protected by some effectual means, they will gnaw nearly every particle of green bark from the body of the tree, for several inches from the ground; and sometimes, when deep snow is piled high around fruit-trees, mice will dig holes through the drifts until they arrive at the trees, and girdle them several feet from the ground. Mice seldom gnaw trees that are not surrounded by a bank of snow at least several inches deep. They never work above the snow, nor girdle a tree that is not surrounded with grass or some other material that will serve as a protection to them while they are committing their depredations. Trees that stand so exposed to the wind that all the snow, grass, leaves, and other rubbish will be swept away, are never in danger of being injured by mice.

The usual means employed to protect trees from mice consists in raising a mound of earth around each tree a foot or more high, as shown by Fig. 68, p. 166. As the marauders work along beneath the grass or snow, the bank of earth turns them away from the tree. When trees are standing in grass-yards or lawns, where the earth is smooth around the bodies, sods may be transported from some other place and fitted closely to the trees, and then removed before the growing season commences. If the snow falls deep around trees, or drifts are formed around the bodies, by stamping the snow all around the trunk of each tree, mice will turn away from the snow that has been packed closely, and not dig a road through to the tree. These remedies have been found effectual, for many years past, wherever the means have been employed. But the labor of packing the snow, when it is deep, around all the trees above the mound of earth is often neglected, to the ruin of many trees.

Another remedy, which has recently been employed with satisfactory success, is winding coarse paper or old cloth

around each tree, securing it with small twine, and afterwards smearing the outside with a coat of coal-tar or pitch.
If the tar were applied to the bark of the trees, no animal
would gnaw the bark. But coal-tar is so poisonous to trees
and plants that, unless the bark is very thick, it will soon
destroy all the vitality of the bark. Two thicknesses of
coarse brown paper will absorb a heavy coat of tar, so that
none of it will reach the bark, nor adhere to any thing that
might come in contact with it. Paper tarred after it has
been put around trees will secure the bark from being
gnawed off by goats, sheep, rabbits, and all other animals.
As soon as the growing season has commenced, the tarred
paper may be removed, although there is no danger that it
will injure a growing tree were it kept around it during
the summer. This manner of protecting trees from almost
every foe will not cost over two to four cents per tree.

Meadow Mice.—There are several species of this kind of mice, some of which are
found in almost every State in the Union. We shall notice only the Prairie-meadow Mouse (Fig. 82, p. 217), the Wood-meadow Mouse (Fig. 83, p. 218), and the
Long-haired Meadow Mouse (Fig. 84, p. 219), the female, and (Fig. 85, p. 220)
the male. Where several species are found in one locality, they are commonly considered by farmers as one animal, known under various names, as Short-
tailed Field Rats or Mice, Bear Mice, Bull-headed Mice, Ground Mice, Bog Mice,
etc., while many persons call them moles, though they are not in the least related
to that family. The food and general habits of the different species are much
alike, though some prefer high and others wet ground; while others inhabit the
woods, prairies, etc. All the species burrow, and none climb trees. Their common food is the grasses and other herbaceous plants, their seeds and roots, and
the seeds and acorns, as well as the bark of trees, in the woods, with grain and
vegetables, when inhabiting cultivated fields. Some kinds, at least, lay up stores
of food for winter. All are active at this time, moving about in the coldest weather, and never hibernate like marmots. One characteristic, certainly, possessed by
all the species in common, is their ability to destroy the products of the farm. In
a nursery, where apple-seeds were planted in autumn, I have observed that, during
fall and spring, so many of the seeds were dug up by these mice as to leave long
gaps in the rows of seedlings, the empty shells of the seeds being found lying
about the rows from which they had been taken. They congregate in stacks of
grain and hay, sometimes in exceedingly great numbers, destroying all the lower
parts, by cutting galleries through them in every direction.
The greatest mischief done by meadow mice is the gnawing of bark from
fruit-trees. The complaints are constant and grievous, throughout the Northern
States, of the destruction of orchard and nursery-trees by the various species of
arvicolæ. The entire damage done by them in this way may be estimated, perhaps, at millions of dollars. This is especially the case at the West, where ▮ care

is taken to protect the trees against them, careless orchardists allowing grass to grow about the roots of their fruit-trees, and thus kindly furnishing the arvicolæ with excellent nesting-places in winter, and rendering the trees doubly liable to be girdled. In the nurseries in Northern Illinois, we have seen whole rows of young apple-trees stripped of their bark for a foot or two above the ground. Thousands of fruit-trees, as well as evergreens and other ornamental trees and shrubs, are at times thus killed in a nursery in one winter. The mice are most mischievous in winters of deep snow. One reason why fruit-trees are most girdled in times of deep snow is, that the meadow mice can then better move about, at a distance from their burrows, being protected by the snow, under which they construct numerous pathways, and are thus enabled to travel comfortably in search of food, always to be obtained in abundance where there is any kind of perennial grass, or the seeds of annual plants. Aided by the snow, too, they climb up at the sides of the trees to gnaw the bark at a considerable height from the ground. Rabbits are often accused of gnawing bark from trees, when the mischief has real-:y been done by meadow mice. All the species alluded to are exceedingly pro-:lific. When unmolested, the female will frequently rear three litters in one season, and each litter will frequently number six or seven young mice. The male Prairie-meadow Mice are always very pugnacious, biting and striking at any thing thrust towards them. When much teased in this way, they sometimes turn on their backs, snap with their teeth, and strike with all four feet. When enraged, they utter a low, harsh, creaking note, resembling that of a very young puppy. If hurt, their voices are clearer and sharper. Sometimes they chatter their teeth in anger.

The males of nearly every species, when they can not escape into some secure

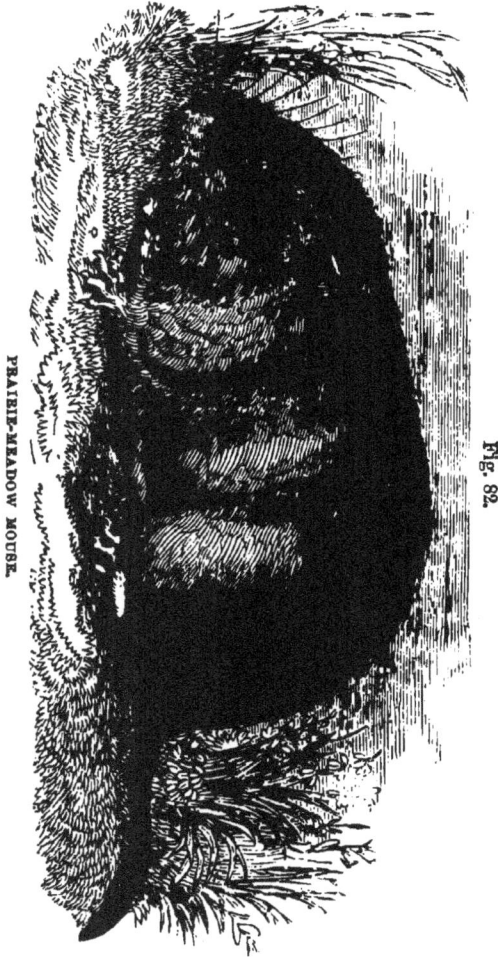

PRAIRIE-MEADOW MOUSE.

Fig. 82.

refuge when attacked by persons, will attack their aggressors with great fierceness. We have often extended the toe of one boot into the fore-front of the battle, when an infuriated mouse was the invading foe, and the plucky little fighter would leave the prints of his sharp teeth in the leather. Thus much for the natural history of the foregoing species of mice. The chief practical consideration is, how to prevent their ravages?

Meadow mice of all kinds have numerous enemies. All kinds of hawks capture a great many. Cats destroy large numbers; and skunks, when they can find a nest of young ones, will devour an entire litter. They often construct their nests on the surface of the ground, in the grass, in stone-heaps, beneath boards and pieces of rails, in old ant-hills, beneath old stumps, and in the ground. Hence, every farmer and gardener should keep several active cats — not old and indolent pussies — but young and fierce mousers, which should be fed well every day with milk. A young female cat is far better than a male; and if the female be allowed to rear two or three kittens, she will destroy large numbers of mice. A well-fed female cat will catch more mice, *when well fed* on milk and fragments from the table, than if she be compelled to subsist wholly on mice. If a cat be allowed to rear two kittens, she will hunt mice a large proportion of the time; whereas, if her kittens be killed soon after their birth, pussy will have but little inclination to hunt mice.

The refuges of mice must be destroyed. Let all the old pieces of rails and bark be placed on end, and let old log-heaps and piles of brush be burned, so that cats and other animals can seize them before they can reach a refuge. When mice are met with in meadows, and when one is ploughing, let every one be killed. When old rail-fences have settled down close to the ground, and the tall grass

Fig. 63.

WOOD-MEADOW MOUSE.

has fallen down in the nooks, mice want no better refuge. Hence, let the fence be lifted with a lever, and a large stone or block of wood placed beneath each corner; and let the dead grass be raked into heaps for manure. Mice frequently burrow among the stones of under-drains, and there build their nests and propagate. In such instances, let old nail-kegs, fish-kegs, old butter-firkins, and barrels be sunk in the ground, so that the top will be even with the surface of the ground. A large number of mice will be trapped in such receptacles, as they will jump in and can not climb out. Boxes two feet deep may also be set in the ground as traps for mice. Cats will take them from such traps, or they can be speared with a fork. It is an excellent practice to sink several barrels in different parts of an orchard. Let two or three holes, one and a half inches in diameter, be bored through the staves about eight inches from the upper end; then let the barrel be sunk in the ground, so that the holes will be even with the surface. The mice will

LONG-HAIRED MEADOW MOUSE—FEMALE.

Fig. 94.

be attracted to the barrels in the night and enter the holes. In addition to all the foregoing suggestions, we have found it an excellent practice to set a few sheaves of corn-stalks in round shocks at different parts of the orchard, beneath which a few nubbins of corn were placed. The tops of the stalks were bound so tightly that the entire shock could be lifted by one man from the ground. In cold weather the nocturnal marauders would leave their haunts and collect beneath the shocks of stalks. Every day we would lift the stalks and let the cat take the victims. There are some other ways of exterminating meadow mice; but none are more convenient and effectual than those alluded to. The war with mice must be waged unremittingly during the entire year. Meadow mice often do great damage in under-

220   *THE APPLE CULTURIST.*

drains. When large numbers come from adjoining woods to a field of grain, or towards an orchard, let several barrels be sunk in the ground, as suggested, between the woods and the orchard. If water settles in the barrels, all the better they will be for mice-traps, if a few leaves and some grass be strewed on the surface of the water.

When an orchard is located near the forest, where mice and rabbits abound, the only reliable means of protecting trees is represented p. 166. Before cold weather has sealed the surface of the ground, a quantity of strong paper should be procured and wrapped around the trees, as already directed. This job should not be neglected, with a view of doing it at some convenient period in the winter. In case a deep snow should fall, mice may ruin half the trees in a single day. If coarse canvas or old sail-cloth can be obtained, such material will be preferable to paper; and if removed with care from the trees, the pieces may be used many years.

In some instances the wood-meadow mice have made a general stampede from a near forest.

Naturalists have described numerous species of field-mice. There are over twenty species of the *Arvicola austerus* (Prairie-meadow Mouse). In many instances owls destroy large numbers. Hawks will spy them from a distance as they soar through the air, and will dart down and swoop up their prey with surprising agility. We have frequently seen a hawk dart from the top of some tree in the open field and swoop up a mouse more than two hundred feet distant from the tree. Snakes also devour more or less of these pests of the orchard and grain-fields. Yet field-mice are so prolific that all their foes united do not exterminate them.

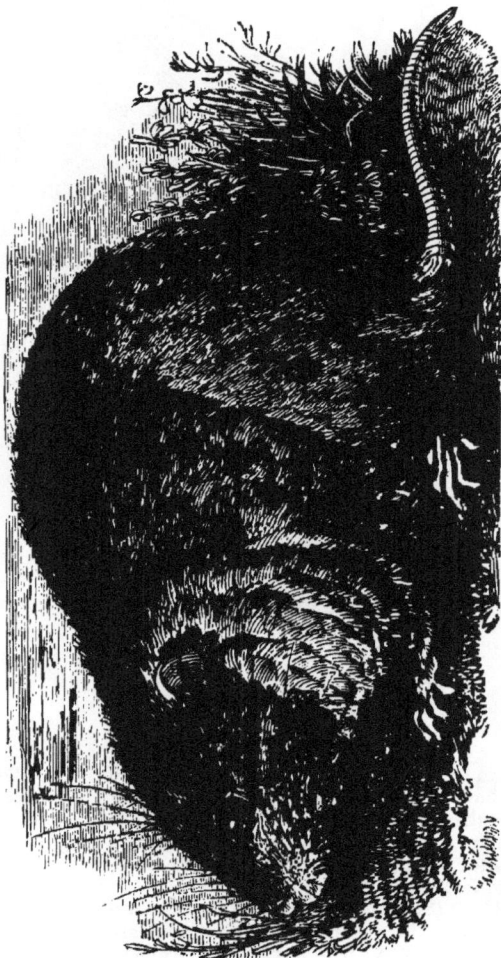

Fig. 85.   LONG-HAIRED MEADOW MOUSE—MALE.

Culture of Apple-trees on Prairies.—A great many persons, who pass for shrewd and intelligent men, do not hesi-
tate to affirm that apples can not be grown with satisfactory success on the Western Prairies; and they will point
to the untold number of failures on the prairies as incon-
trovertible evidence that their assumptions are reliable.
Yet our doctrine is, that bountiful crops of apples can be
grown on any ground that will yield fair crops of wheat,
provided the trees are produced and cultivated as directed
in the former part of this book.

Dr. J. A. Warden, who is accepted authority at the
West, states, in the "Prairie Farmer Annual," that it has
been too much the custom, in many portions of the West,
to accept as true the oft-repeated dogma that orchards can
not be grown on the prairies. It is surprising that such a
notion should still prevail among many of our large farmers, since the contrary has been so fully proved in every
county, and in almost every township. The failures, in almost every locality, may be traced to some, or to all of the
following causes: 1st. The bad condition of the trees planted; 2d. The bad selection of varieties; 3d. The want of
suitable preparation of the soil; 4th. Bad planting, and
want of suitable cultivation and training; 5th. Exposure
to the elements, and to the depredations of stock and wild
animals.

This bad condition of the trees may arise from the dangers or delays of long transportation; from carelessness in
handling and defective packing; from frosts, when exposed
to the air; or simply from drying of the roots, when care-
lessly exposed, after they have been received. It is unfortunately true, also, that some nursery-trees are badly grown,
especially when they have been much crowded, and when
the nurseryman has injudiciously sacrificed stockiness for
height. Thus the blame is distributed all the way from a

proper selection of the apple-seeds to the final operation in
the management of the bearing trees.   Bad selection of varieties has been a frequent source of failure in attempts to
establish Western orchards.   It was most natural for men
to desire, in their new homes, the same fruits to which they
had been accustomed in their old ones.   Hence, they introduced chiefly the varieties of New England, many of which
have proved worthless on the prairies; and, strange to say,
some of those which have suffered most from winter-killing
originated in the Eastern States; while some of the most
hardy varieties have come from the South.   The labors of
Western pomologists have at length enabled us to select
lists of hardy varieties that have withstood the severe or-
deals of such winters as those of 1856 and 1863.

The want of suitable preparation of the soil has been a
too common explanation of the failures complained of, particularly the want of drainage in low and flat lands, with a
retentive subsoil.   This is a matter which can be soon
overcome with the progress of agriculture.   The opening
of the country, the introduction of thorough drainage, and
the perfecting of the surface-drainage, will enable farmers
to grow orchards where they have previously failed.   People must learn that the ground must be well prepared for
the orchard, as for any other farm crop; that the trees
need cultivation, especially while they are young; and when
older, that they require the sole occupancy of the ground.
At the same time, it will be generally understood that Nature should be aided and directed in the formation of the
trees by judicious training and pruning.   Trees planted out
in the open prairies, exposed to the winds, sunshine, and
storms, without any protection, must be expected to suffer
in the change from the sheltered nursery-plat on which they
were produced, and that, perhaps, in a different soil and climate, certainly under very different conditions from those

of the open field. If, in addition to this, they be so neglect-ed as to suffer from the attacks of mice, gophers, or rab-bits, and to the inroads of animals of larger growth, so as to be barked by mules or sheep, and browsed upon or broken down by horned cattle, we need not wonder at the state-ments that have been made regarding the impossibility of growing orchards in the prairie.

Another prominent peculiarity in the lands appropriated to orcharding in the Western country consists in their gen eral flatness. This necessitates their open exposure to the winds, which sweep with great force across the plains; and when the dew-point is low, evaporation from all living tis-sues is greatly increased, which causes the trees to dry away fearfully. The same flatness of the land favors the accumulation of surface-water, which makes the ground wet and clammy, requiring judicious efforts to effect suit-able drainage over extensive tracts of country. It is emi-nently important that such flat land, where the subsoil is of a retentive character, should be thoroughly under-drained. Under-draining, in many instances, is the only requirement to insure the success of apple culture.

When the land is very flat, if it has not been under-drained, it will be well to plough it in lands, gathering the furrows repeatedly on the same line, so as to ridge the sur-face as high as possible where the trees are to stand, so that the dead-furrows will make an open ditch for the water, and the gather-furrows raise the roots of the trees' above general water-level. These ridges and furrows should be made in the direction of the slope of the natural surface, if any can be detected; otherwise the ditches must be made deeper at one end than the other. It is rarely the case that the surface-water will not find an outflow at one end or the other of a ten-acre field. Towards this the ridges and fur-rows should extend. Besides the foregoing considerations,

orchards on the prairies must be shielded by belts of evergreens during the severity of winter. By heeding all the
suggestions alluded to in the preceding paragraphs, and
particularly the directions laid down in Chapter I., a failure
of an apple-orchard will be of rare occurrence. :9''

Cause of Barrenness in Apple-trees.—Barrenness and indifferent productiveness of apple-trees arise from numerous
causes. In most instances, the difficulty is in the barren
soil rather than in a tree. Yet, when a part of the trees
are productive, the evidence is conclusive that the fault is
in the tree. In many instances, some of the largest trees
in the orchard rarely yield a bushel of fruit, while those on
either side of them yield abundant crops. Dr. Hull, an
eminent pomologist, states that apple-trees that expend all
their forces in the production of wood-growth can produce
little or no fruit. Indeed, it is not possible for any tree to
produce a fruit germ, and not again in some way disorganize it, unless the wood-growth shall cease in time for
the leaves to elaborate food enough to grow both leaf and
fruit the following year, or until a part of the leaves shall
attain to nearly or quite their full size. That this is so, will
be apparent when we consider that the leaves which first
appear in the spring were formed in the bud the previous
year, perfect in all their parts, and in the embryo state contained each individual cell found in them when fully grown.
In order to produce an abundant crop of fruit, the extension of the shoots must cease a long time before the end of
the growing season, so that the wood may mature and the
fruit-buds develop.

The question may arise, if there is no addition to the
number of cells, how do the leaves grow? The answer is,
that the only difference we can see, between an embryo leaf
and one fully grown, is in the size of the leaf-cells. As
growth begins in the spring, these small cells, which were

formed in the previous year, begin to expand. Each individual cell thus enlarges, until all the numerous cells of which these leaves are composed are of full size. To further illustrate this, let us suppose, in a brick wall, that each brick at the same time was gradually to expand to several hundred times its present diameter, and you have just what takes place in the growth of an embryo leaf. Here we have a tree in possession of a full-grown leaf. This leaf did not form itself, but was formed by the tree in the preceding year. To produce and sustain this cellular enlargement, there had been stored, the previous year, a large share of nutriment in the buds and in other parts of the tree. This nutriment must not only be sufficient to feed the embryo leaves, but must also be sufficient to produce the small warty excrescences — the rootlets and spongioles. These new leaves and spongioles are a tree's laboratory. And those leaves and spongioles first grown were made, with the exception of moisture, wholly out of the materials that were stored by the tree during the growth of the previous year. When these vegetable stores are in sufficient supply to do this, and nourish the fruit germs also, then we shall hear little about imperfect fertilization. On the other hand, had the food been consumed the previous year, by ripening an over-crop of fruit, or by making a very succulent growth, then the tree would not store a sufficient amount of plant-food to perform its threefold office in the production of leaves, roots with their spongioles, and fruit. In this condition, a part of the leaf, and a larger part of all the fruit-buds, yield up their nourishment, which goes to the production of root and leaf growth. The tree, therefore, is barren of fruit for the summer, its whole growth being required to recuperate its own vigor. Such trees often bloom freely, and then cast their blossoms. When this occurs, uninformed persons often attribute its barrenness to want of fertilization, or suppose that the rain has washed away the pollen. When an apple-tree ripens its wood early in the season, and ceases to grow late in autumn, if the soil is fertile, the fault is in the tree; and the sooner it is removed, the more satisfactory it will be for the proprietor. Some trees are naturally poor bearers, or no bearers at all, and they can never be made to yield a crop of fruit.

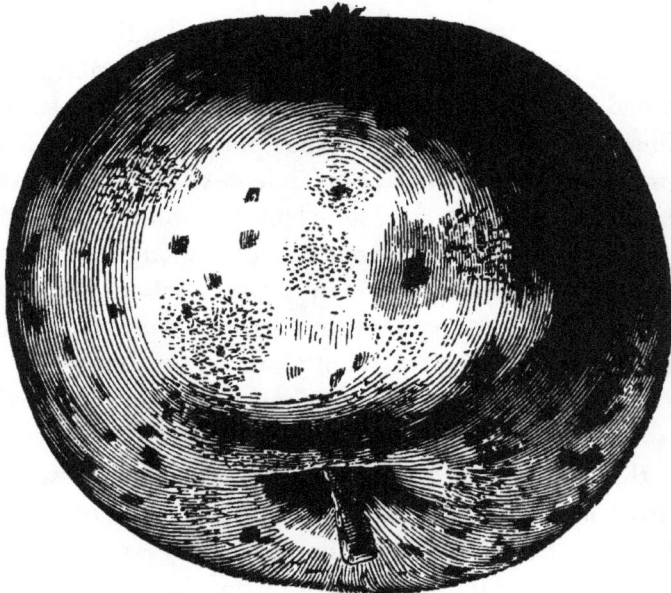

FENOUILLET JAUNE.

Synonyms.—Drap d'Or, Pomme de Charactère, Embroidered Pippin, and Cloth of Gold. This beautiful apple is of French origin. The fruit is covered usually with a yellow gray russety net-work. The stem is beautifully dotted with small dark specks; slender, and in some specimens short and stout. The basin is shallow; taste, sub-acid, and fine-flavored. Season, early autumn, continuing until cold weather; tree, vigorous, regular, and spreading, and a prolific bearer. This beautiful variety has not as yet been introduced and cultivated except by a small number of pomological amateurs. We have never seen the fruit in the New York market. A few nurserymen who cultivate rare varieties, have grafted a few trees with cions taken from a "Cloth of Gold" tree. By communicating with gentlemen who cultivate extensive nurseries near New York city, or Rochester, or Chicago, Ill., a few cions or young trees may readily be obtained. By means of superior cultivation, the size may be greatly augmented.

CHAPTER VIII.

EXTERMINATING NOXIOUS INSECTS FROM APPLE-ORCHARDS.

The army-worm, canker-worm, palmer-worm, slugs,
The joint-worm, the drop-worm, the borer, and bugs,
Like rapacious armies, when seizing their prey,
United, spread ruin through branches and spray.—EDWARDS.

The Increase of noxious Insects.—It has been affirmed by some pomologists and tillers of the soil that in proportion as we increase improved fruits, just in that proportion will fruit-insects, and fruit and fruit-tree diseases increase. A recognition of this fact will each year, as we multiply orchards, become more and more apparent. Hale's Early peaches, at first, will be free from rot. Pear-trees will be measurably exempt from pear-tree blight; vines free from Vine-hoppers, and grapes free from Grape-codlings and rot. From some cause not yet well understood, all, or nearly all, our young vineyards are, for the first few years of fruitage, free from rot; and then ever afterwards subject to it. The same is true of cherry, peach, and plum rot. Therefore, to those engaging in pomological pursuits, a knowledge of the several difficulties likely to be encountered should be recognized; and, so far as known, the remedies for each difficulty must be promptly applied. For many years past, only a small proportion of the people who attempt to grow fruit have been careful to exterminate one-fourth part of the noxious worms and insects on their premises. Hence they have multiplied with surprising rapidity. Consequently, as orchards and crops of fruit have increased, and insects have been allowed to flourish unmolested where

they found congenial quarters, it has frequently appeared
that the increase has been without a parallel. But noxious
insects are sometimes an incentive to prompt indifferent
pomologists to dispute their right to the growing fruit.
The "American Entomologist" affirms that more than two
dozen species of noxious insects—among which the varieties are unknown, as to numbers—are now met with as
formidable depredators in the apple-orchards of many of
the States. Some subsist upon the root; some burrow into
the trunk; some infest the bark; some select the opening
buds; some devour the expanded foliage; and others, final-
ly, revel upon the fruit. Thus beset by enemies on every
side, it would seem that the good old apple-tree must ere
long succumb, and cease to occupy its place in the family
of plants; and this would undoubtedly be the case if all
these enemies were permitted to go on unchecked in their
operations. But owing to the incessant antagonism of
parasitic foes, insectivorous birds, and human ingenuity,
the ravages of these insects are kept within bounds, and
the apple-tree still lives. Of these numerous enemies of
the apple-tree, five hold a bad pre-eminence—namely, the
Round-headed Borer (*Saperda bivittata*), the Oyster-shell
Bark-louse (*Aspidiotus conchiformis*), the Canker-worm
(*Anisopteryx vernata*), the Tent-caterpillar (*Clisiocampa
Americana*), and the Apple-worm (*Carpocapsa pomonella*).
Of these the most conspicuous, and in some seasons and localities the most destructive, is the insect generally known
as the Tent-caterpillar (Fig. 95, p. 253).

The true Way to exterminate Depredators.—We might
swell this volume to hundreds of pages touching the natural history, habits, and metamorphoses of noxious insects,
worms, and caterpillars, all of which would be interesting
and instructive; but the great practical consideration is,
will such a fund of information render any practical aid in

saving fruit and fruit-trees from the ravages of such depredators? Well-conducted experiments have been made so
frequently, to repel insects, worms, and caterpillars, that it
is safe to assume that fruit of nearly every kind must be
saved, while growing, by active vigilance on the part of pomologists. New depredators have been visiting our fruit-
trees every season for a number of years past; and for years
to come others, now unknown, will probably appear. There
may be some remedy discovered to head them off; but the
most reliable one of all will be to "catch 'em and kill 'em."
It would seem, in many instances, as if the caterpillars, the
Army-worm, the Canker-worm, and numerous other depredators that subsist on the leaves of apple-trees, would cut
off the entire apple-crop, if we do not "catch 'em and kill
'em." The Tent-caterpillar, in many instances, is "doing
his level best" to devour every leaf on apple and cherry
trees. There is no possible way to prevent their ravages,
unless we "catch 'em and kill 'em." Entomologists may pen
interesting paragraphs about the origin, hibernation, metamorphoses, and the habits of noxious insects and worms,
and suggest nostrums to repel them and check their ravages; but, after all that may be said or written, if we save
our trees and fruit, in most instances we must "catch 'em
and kill 'em." Apple-tree Borers, Peach-tree Borers, Cur-
rant Borers, Pear-tree Borers, and borers for almost every
tree and plant that dares to grow, are incessantly working
their gouge-shaped augers night and day. Hence, the only
watch-word should be: "Catch 'em and kill 'em." All
through the growing season, every employé on the premises should be instructed, wherever he sees a noxious insect
at work, to drop all other employment, and "catch 'em and
kill 'em." We have tried the "shoo-fly" remedy quite too
long, without any satisfactory results. If we fray them
away, they are back to their work of devastation before we

can return.   If we "catch 'em and kill 'em," they never
have a resurrection.

Near neighbors, when insects or depredators of any kind
are numerous, should all labor in concert to exterminate
noxious invaders.   So long as one or two persons, having
orchards, neglect to destroy depredators on their ground,
vigilant pomologists may put forth their efforts in vain to
save their fruit from the ravages of the curculio and caterpillars.   At Vineland, New Jersey, all the inhabitants of
the town, of 12,000 people, where they have heretofore produced the finest fruit in the country, rely on no other remedy than to "catch 'em and kill 'em."   We may depend
on the efficacy of hellebore to destroy the Currant-worms;
but as yet, no reliable means have been found to keep the
hordes of insects from destroying the leaves, the blossoms,
and the fruit of cherry-trees, plum, peach, and apple trees,
except this—"catch 'em and kill 'em."   Every pomologist
should instruct those in his employ to keep a watchful eye
for all manner of depredators, and to "catch 'em and kill
'em" as quickly as they would dispatch a rattlesnake.   The
price of fruit is unremitting vigilance and war to the knife,
with the knife to the hilt.   Numerous other modes have
been strongly recommended, such as covering the trees
with lime-wash or tobacco-water, smoking the trees daily,
placing putrid substances under them, spading-in the rising
curculios, cutting canals under the trees to fill with water,
laying brick pavements, making mortar floors, and other
modes hard to apply, and of little or no efficiency.

Exterminating Moths, Millers, Flies.—A large proportion
of insect enemies originate from millers or flies, many of
which are seen on the wing at evening, or at any time at
night.   Most persons understand how common it is for millers to fly round about a lighted lamp, and frequently dart
directly into the flame.   We have often known doves and

swallows to fly down to our barn-lamp when we have been about the out-buildings. This fact will often enable pomologists to destroy insect enemies by thousands simply by setting a lighted lamp or candle within a glass jar, or beneath a bell-glass, so as to attract millers to the light. But all such glasses must be open both at the bottom and top, as a short piece of lighted candle let down to the bottom of a tight glass jar will burn only for a few minutes. The burning wick must be supplied with fresh air. A common barn-lamp, set in the centre of a large milk-pan containing a quart or more of water, sweetened thoroughly with cheap molasses, will often constitute a trap that will attract and decoy into the sweet liquid a large number of millers, which would scatter an untold number of eggs, every one of which would produce a worm or caterpillar that would destroy more or less fruit. It has been frequently recommended to kindle small fires in different parts of an orchard with shavings or straw. In many instances, millers will fly directly into the burning fire, and be destroyed at once by the flames. (See page 268.)

Some writers have recommended oil instead of sweetened water. Others have suggested that the bell-glass or globe of the burning lamp be smeared with oil or molasses, so that the nocturnal invaders may stick fast in the glutinous material when they fly near the light. Downing alludes to an experiment of the kind in France, during which it was stated that 30,000,000 millers were destroyed in one vineyard. As this is a large number, we must make proper allowance for the great distance this story has travelled. Doubtless, during such a long and perilous journey, as many as two right-hand ciphers have been accumulated. Nevertheless, we can recommend this mode of destroying millers in orchards and vineyards; and melon-bugs may be exterminated, not in myriads, as Downing affirms, but in large numbers. When millers fly into our dwelling, and flutter about the lamp in the evening, by destroying them a large number of caterpillars or other noxious depredators may be exterminated.

Bottle-traps for Insects.—One of the most convenient and efficient ways to trap many kinds of insects, bugs, beetles, and millers, is to procure a large number of small bottles, and fill each about half full of switchel, adding a few spoonfuls of cider-vinegar to a quart. Then make little shelves, and nail them to the side and limbs of the trees, as shown in Fig. 51, page 117, for receiving the bottles. The insects will find this aromatic beverage, and will collect in large numbers within the bottles, so that the contents will have to be changed sometimes every day. Some writers have recommended to suspend such bottles by a cord. But it will be found a much more satisfactory way to place the bottles on small shelves, as many insects will not climb down a cord to a swaying bottle. We feel confident that a person will never suspend his bottles more than once if he can procure such little shelves as we have recommended.

The Canker-worm (*Anisopteryx vernata*).—This pest of our fruit-orchards is distinguished from most other caterpillars that attack the apple, by having but four prolegs at the end of the body. The normal number of such prolegs in caterpillars is ten; and it is the lack of the foremost six which obliges this insect to span or loop, from which habit the characteristic name *Geometridæ* has been given to the family to which it belongs. The Canker-worm is by no means confined, in its destructive work, to the apple; for it likewise attacks the plum, the cherry, the elm, and a variety of other trees. The full-grown worms vary greatly in the depth of shading and in the ground-color—different shades of ash-gray, green, and yellow, almost always occurring in different individuals of the same brood. This same variation in color is common to most other span-worms. Professor Riley once stated that the word *canker-worm* has formed the heading of so many articles in Agricultural and Horticultural journals, during the last ten or twelve years, and its natural history has been so fully given in the standard work of Dr. Harris, that one almost wonders where there can be a reading farmer who does not know how properly to fight it. But we must remember that new generations are ever replacing those which pass away, so that the same stories will doubtless have to be repeated to the end of time. Facts in Nature will always bear repeating. Hence, as it may be laid down as a maxim that no injurious insect can be successfully combated without a thorough knowledge of its habits and transformations, we will first recount those of the Canker-worm, and afterwards state the proper mode of extermination.

The figures herewith given represent a full-grown Canker-worm, and the same pest in its different stages. The

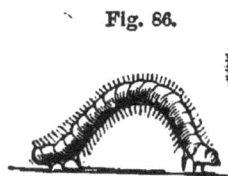

Fig. 86.

Fig. 87.

Fig. 88.

Larvæ of the Canker-worm.

Miller of the Canker-worm.

Female moth of the Canker-worm.

eggs of this insect are very minute, short, and cylindrical, and are deposited close together in rows, forming batches. They are glued together by a grayish varnish which the mother moth secretes, and they are attached to the trunk, or to some one or other of the twigs of the tree, each batch consisting of upwards of a hundred eggs. As the leaves begin to form, these eggs hatch into minute, thread-like span-worms; and in from three to four weeks afterwards

they acquire their full size, when they appear like Fig. 86.
After the Canker-worm has attained its full size, it either
crawls down the tree or lets itself down by means of a silk-
en thread, and burrows in the ground, where, at a depth of
two or three inches, it forms a rude cocoon of particles of
earth intermixed with hair. Within two days after com-
pleting the cocoon, the worm becomes a chrysalis of a light-
brown color. The sexes are now distinguishable, the male
miller, shown at Fig. 87, being slender; while that of the
female is much larger, and destitute of wing-sheaths (Fig.
88). In the New England States, the worms descend into
the ground during the last of June, where most of them
remain till early the following spring, though many of them
change to moths and issue during the mild days of early
winter.

Manner of Extermination.—The sole object of the female
moth of the Canker-worm, after she leaves the earth, seems
to be to provide for the continuance of her kind; and she
instinctively places the precious burden, which is to give
birth to the young which she herself is destined never to
behold, upon the tree whose leaves are to nourish those
young. ·All her life-energy is centred in the accomplishment of this one object; and she immediately makes for
the tree upon issuing from the ground. Consequently,
any thing that will prevent her ascending the trunk will, in
a great measure, preserve the tree from the ravages of the
worm. Numerous, indeed, have been the devices to accomplish this desired object. Tar, applied to strips of old canvas, sheep-skin, or stiff paper; refuse sorghum-molasses, or
printers' ink, applied in a similar manner; tin, lead, and rub-
ber troughs to contain oil, tin-plate collars sloping downward, belts of cotton-wool, etc., have all been used with
satisfactory results, in most instances, according as they
have been used intelligently or otherwise. All these ap-

pliances, of whatsoever character, are divisible into two classes; first, those which prevent the ascension of the moth by entangling her feet and trapping her fast, or by drowning her; and, second, those which accomplish the same end by preventing her getting a foothold, and thus causing her repeatedly to fall to the ground, until she becomes exhausted and dies. The most economical, efficient, and least troublesome manner of exterminating this pest is to provide numerous bottles of sweetened water, as represented by Fig. 51, p. 117. If two or three such bottles be placed near the foot of each tree, almost every miller will be trapped. The orchardist must know that many of the moths issue in the fall of the year, and that the applications must be made at least as early as the former part of October; and that the tarred bandages must be kept sticky, through all mild weather, till the leaves have appeared in the following spring. Furthermore, he must know that many of the moths—frustrated in their efforts to climb the tree—will deposit their eggs near the ground, or anywhere below the tar; and that the young worms hatching from them are able to pass through a small crevice, or over a fine straw. Thus, if troughs are used, they must be fitted over a bandage of cotton-wool, so that when the trough is drawn tightly around the tree it will do no injury, and will, at the same time, cause the cotton to fill up all inequalities of the bark. The joint must likewise be kept smeared either with tar or molasses, and then the worms will not be able to pass. In the neglect to thus fasten them lies the secret of failure which many report who use such troughs. Young worms can march over the smoothest glass by the aid of the glutinous silken thread which they are able to spin from the very moment they are born.

It will require much persevering effort, time, labor, and expense to continually renew the applications of tar on every tree in a large orchard during so many months of the year. But bottles of sweetened water may be quickly renewed after the contents have been cast out. The old adage, " What is worth doing at all, is worth doing well," was never truer than in fighting this insect. Apply the remedy thoroughly during two successive years, and you will have utterly routed the enemy; and this is more especially the case where an orchard is not in too close proximity to a forest or to slovenly neighbors. Fail to apply the remedy, and the enemy will, in all probability, rout you. The reason is simple. The female being wingless, the insect is very local in its attacks, sometimes swarming in one orchard, and being unknown in another which is but a mile away. Thus, after it is once exterminated, a sudden invasion is not to be expected, as in the case of the Tent-caterpillar, and of many other orchard pests ; but when the Canker-worm has once obtained a footing in an orchard, it multiplies the more rapidly, for the very reason that it does not spread fast. As it is always easier to prevent than to cure, it were well for the owners of young orchards, in neighborhoods where the Canker-worm is known to exist, to keep a sharp look-out for it, so that upon its first appearance the evil may be nipped in the bud. In the same manner that it is exterminated in the individual orchard, it may, by concert of action, be exterminated from any given locality. When once the worms are on a tree, a good jarring will suspend them all in mid-air, when the best way to kill them is by swinging a stick above them, which breaks the web, and causes them to fall to the ground, when they may be prevented from ascending the tree by the tin flange around the tree (Fig. 51, p. 117). Birds will destroy a large number of these pests, when they are young and tender, in the larva state. There will be no difficulty in exterminating the Canker-worm from every orchard, if proper efforts are employed at the right season of the year. The full-grown larvæ must be crushed by hand or with the insect-crushers (Fig. 89), which may be made of wood, or be purchased at hardware stores.

. **The Curculio** (*Rhynchœnus nenuphar*), "Little Turk."— The great mass of mankind are comparatively ignorant with regard to the real character and habits of this depredator. We pretend to know but little about it. And yet many who assume to be familiar with what is called

Fig. 89.

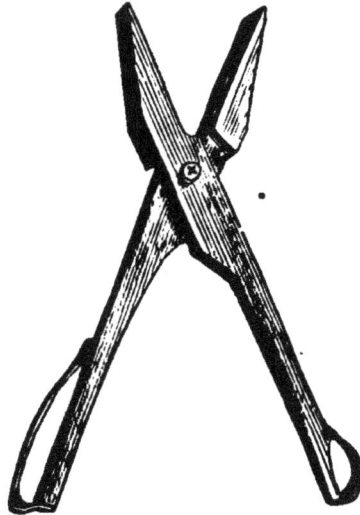

Insect-crushers, or pincers.

the "Curculio," do not know any more about it than the writer. We have the Apple Curculio and the Plum Curculio, and several other kinds of beetles called "the Curculio." But we have searched in vain through all our horticultural and pomological journals to find a description or an allusion to the curculio, in which the difference between the two is pointed out. So much has been written on the habits of this one little insect, and on the best means of protecting our fruits from its injurious work, that one almost tires of repeating those established facts in its history which all who are interested should know. There are yet some mooted points to be settled in the natural history of our curculio. There are actually many fruit-growers who do not know a curculio when they see one. The curculio is such a formidable enemy to our choice fruit, that every country boy ought to be taught to understand the character of this depredator, so that he may not be in doubt when he meets with one, any more than when he sees a tent-caterpillar, as to the name by which the foe may be called.

We herewith give life-like illustrations, from the " American Entomologist," of both the Plum Curculio and the Apple Curculio. Figure 90 is a magnified representation of the Plum Curculio in its various stages. The magnified grub at *a* represents the curculio in the *larva* state. The horizontal line beneath it represents the natural length when it is found in the fruit. At *b* the insect is represented in the *pupa* state. The vertical line on the left side of it shows the natural length. This *pupa*, or *chrysalis*, is greatly magnified. Its color

Fig. 90.

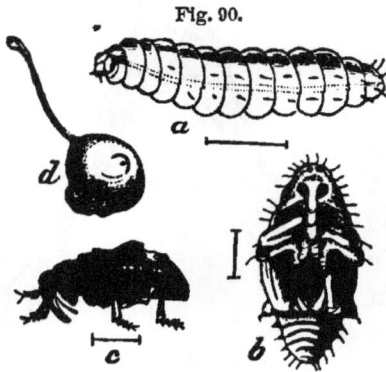

The Plum Curculio.

a, magnified larvæ; *b*, the chrysalis; *c*, a magnified beetle; *d*, a full-sized insect on a plum.

is that of a whitish grub. At *c* the perfect insect is represented as magnified several times. The mark beneath the illustration represents the usual length of a full-grown insect. The color of this " little Turk " is sometimes brown, black, or of a clay-yellow. At *d* a curculio of full size is shown on the surface of a small plum, in one side of which the crescent mark of the curculio's proboscis appears, such as is made on the young fruit when an egg is deposited beneath the skin. Fig. 91, on the opposite page, represents the Apple Curculio, which, to appearance, is a very different insect from the Plum Curculio. At *a* is a diagram of the depredator of natural size. At *b* is a magnified side-view. At *c* we have a back view of the same insect. The usual color of this insect is a dingy gray, inclining to red rust behind. The Apple Curculio spreads

with extraordinary rapidity, and destroys the greater part of the apple crop in
many localities. The Ap-
ple Curculio makes a round
cut difficult to see with the
eye. The worm remains
where the egg was laid
until it matures, when it
comes out and goes into the
ground. The curculio can

Fig. 91.

The Apple Curculio.

not fly under a temperature of 70 degrees. They fly against
the wind; but as yet no one has been able to determine the
extent to which they migrate. The Apple or Four-humped
Curculio (*Anthonomus quadrigibbus*) is a much smaller insect, with a snout which sticks out more or less horizontal-
ly and can not be folded under, *and which is as long as the
whole body*. This insect has narrow shoulders and is broader behind, where it is furnished with four very conspicuous
humps, from which it takes its name. It has neither the
polished black elevations nor the pale band of the Plum
Curculio. In short, it differs generically, *and does not attack the peach*. What we have to record in this connection respecting the curculio will be chiefly of a practical
character.

The Plum Curculio, commonly known all over the country as "the curculio," is a small, roughened, warty, brown-
ish beetle, belonging to a very extensive family known as
Snout-beetles (*Curculionidæ*). It measures about one-fifth
of an inch in length, exclusive of the snout, and may be
distinguished from all other North American Snout-beetles by having an elongate, knife-edged hump, resembling
a piece of black sealing-wax, on the middle of each wing-
case, behind which humps there is a broad clay-yellow

band, with more or less white in its middle. The snout of the Plum Curculio is *scarcely as long as the head and thorax together*, and can be folded back between the legs, where there is a groove to receive it. The Plum Curculio is broadest across the shoulders and narrows behind; and, moreover, the black sealing-wax-like, knife-edged elevations on the back, with the pale band behind them, characterize it at once from all our other fruit-boring snout-beetles. It is in this hard, shelly, beetle state, that the female passes the winter, sheltering under the shingles of houses, under the old bark of both forest and fruit trees, under logs, and in rubbish of all kinds. As spring approaches, the insect awakens from its lethargy; and, if it has slept in the forest, it instinctively searches for the nearest orchard. In Central Illinois and in Central Missouri the beetles may be found in the trees during the last half of April; but in the extreme southern part of Illinois they appear about two weeks earlier; while in the extreme northern part of the same State they are fully two weeks later. Thus, in the single State of Illinois, there is a difference of about one month in the time of the curculio's first appearance on fruit-trees; and the time will vary with the forwardness or lateness of the season.

How they Oviposit.—The curculio first commences to puncture fruit when the specimens are of the size of small marbles, or of hazel-nuts, though she may be found on your trees as soon as they are in blossom. Alighting on a small peach, she takes a strong hold of it, and with the minute jaws at the end of her snout makes a small cut just through the skin of the fruit. She then runs the snout slantingly under the skin, to the depth of one-sixteenth of an inch, and moves it back and forth until the cavity is large enough to receive the egg it is to retain. Then she turns around and drops an egg into the mouth of the cavi-

ty. After this is accomplished, she resumes her first position, and by means of her snout pushes the egg to the end
of the passage; and afterwards deliberately cuts the crescent in front of the hole, so as to undermine the egg and
leave it in a sort of flap. The whole operation requires
about five minutes; and her object in cutting the crescent
is evidently to deaden the flap, so as to prevent the growing fruit from crushing the egg. Now that she has completed this task, and has gone off to perform a similar operation on some other fruit, let us from day to day watch
the egg which we have just seen deposited, and learn in
what manner it develops into a curculio like the parent
which produced it, remembering that the life and habits of
this one individual are illustrative of those of every Plum
Curculio that ever had, or that ever will have, an existence.
We shall find that the egg is oval and of a pearly-white
color. Should the weather be warm and genial, this egg
will hatch in from four to five days; but, if cold and unpleasant, the hatching will not take place for a week, or
even longer. Eventually, however, there hatches from the
egg a soft, tiny, footless grub, with a horny head; and this
grub immediately commences to feed upon the green flesh
of the fruit, boring a tortuous path as it proceeds. It riots
in the fruit, working by preference around the stone, for
from three to five weeks, the period varying according to
various controlling influences. The fruit containing this
grub does not, in the majority of instances, mature, but
falls prematurely to the ground, generally, before the grub
is quite full-grown. We have known fruit to lie on the
ground for upward of two weeks before the grub left,
and have found as many as five grubs in a single peach
which had been on the ground for several days. And
yet it is an unusual occurrence to find more than one or
two grubs in a plum. Some entomologists have stated

that curculios rarely deposit more than *one* egg in a plum or apple.

When the grub has become full-grown, however, it forsakes the fruit which it has ruined, and burrows from four to six inches in the ground. At this time it is of a glassy, yellowish-white color, though it usually partakes of the color of the fruit-flesh on which it was feeding. It is about two-fifths of an inch long, with the head light-brown. In the ground, by turning round and round, it compresses the earth on all sides until it has formed a smooth and oval cavity. Within this cavity, in the course of a few days, it assumes the pupa form, as represented at *b*, Fig. 90, p. 238. After remaining in the ground in this state for about three weeks, it becomes a beetle, which, though soft and uniformly reddish at first, soon assumes its natural colors; and, when its several parts are sufficiently hardened, works through the soil to the light of day.

"**Playing Possum.**" — The curculio, when alarmed, like many other insects, and especially such as belong to the same great order of beetles (*Coleoptera*), folds up its legs close to the body, turns its snout under into a groove which receives it, and drops to the ground. In doing this, it feigns death, so as to escape from threatened danger, and does in reality greatly resemble a dried fruit-bud. It attacks, either for purposes of propagation or for food, the nectarine, plum, apricot, peach, cherry, apple, pear, and quince, preferring them in the order of their naming. It is always most numerous, in the early part of the season, on the outside of those orchards that are surrounded with timber. It is also more numerous in timbered regions than on the prairie. We believe all entomologists agree, touching the foregoing habits and natural history of both the Plum Curculio and the Apple Curculio. As already stated, when this insect is alarmed, it gathers up its legs, drops to the ground, and feigns death. This peculiarity gives us the only effectual method of destroying it, which is to jar the trees every morning, and catch the insects upon a sheet and destroy them. All fallen fruit must be gathered and destroyed; or hogs should be turned into the orchard and allowed to consume every fallen apple or other fruit, while the larvæ are still in the fruit. The warmth that brings out blossoms brings the curculios to their natural food and breeding-places. They hide anywhere in the orchard where there is a cover. During the warm days and nights of April and May they go up the tree—mostly crawling, it is presumed—to feed and to pair.

How to Trap the Curculio.—No possible way has as yet been discovered to prevent the ravages of the curculio, except the summary way of dealing with most other insects, worms, and caterpillars—"to catch 'em and kill 'em." Some pomologists have reported excellent success in their efforts to trap the Little Turk in the following manner. On the contrary, others have attempted the same thing, and reported no satisfactory results. Early in the spring, let the surface of the ground round about every tree be rendered as smooth as practicable by removing every sod and stone, and crushing every clod. Make the ground very smooth around the base of the tree. Do not leave a single hole next the tree. Leave no place where the curculio can hide, except under the shelter you provide. *Now lay near to the tree, and close to the ground,* about four pieces to a tree, either chip, or bark, board, lath, rag, corn-cob, old leather, or any thing for a covert. The curculio will conceal itself under this shelter, and may be destroyed by the thousand. Go around every day and turn over each chip, killing every curculio. They will generally adhere to the chip, but may often be found on the ground under the chip. The first warm day in spring that starts vegetable life calls the curculio forth, when it proceeds to its feeding and breeding ground. They walk fast, and they fly and feed generally at night, eating the young and tender leaves. Except in cold and stormy nights, the curculio, in most instances, remains on the trees to feed, and some pomologists will insist that they usually come down towards morning to hide. They crawl, on cold days and nights, and hide under the shelter of the trunk of the tree, waiting to feed when the nights become sufficiently warm. The curculio uses the green fruit only to hold its egg and young. When the weather is warm and pleasant, almost every insect remains in ·the tree-top. We have seen reports in which the writers stated that they had trapped thousands of them every day beneath their peach-trees, plum-trees, and other fruit-trees. The curculio may often be found on the under side of the most lateral branches, where there is knot, black moss, bud, twig, rough bark, or any thing that will give a partial hiding-place. It is important to be ready for them when they first appear out of the ground in the spring, as they take refuge under old matted leaves, sod, lumps of dirt, sticks, any thing on and around the trees, especially old rough bark, wherever there is a sufficient covert. They move but little at first, unless it is warm. A Western pomologist, W. B. Ransom, wrote that he killed nearly eighteen thousand in a few days; and a neighbor destroyed twenty thousand in his orchards by simply smoothing the surface of the ground, and placing a few chips on the ground near the bodies of the trees. Old pieces of woollen cloth laid in the fork of the limbs will furnish an excellent refuge for them during cold and windy days and nights.

Jarring the Trees to catch the Curculio.—Up to this time of writing, it is generally admitted that there is no mode of destroying the Little Turk that can be relied on as thoroughly efficient, except the practice of jarring the trees and catching the curculios as they fall on sheets spread beneath the trees. The trees are jarred by means of a sharp blow or two in quick succession, which induces every insect to relinquish his hold on the fruit or branch, when they all fall to the ground. Shaking a tree has very little efficacy to make them quit their hold.

The most convenient apparatus, in the line of sheets, consists of two pieces of strong factory, about seven feet wide by fourteen feet long, each one having a half-circle cut out of one side near the middle, to fit around the body of the tree. The sheets may be made of good bed-ticking or of old sail-cloth. As they are spread out beneath a tree, the operator steps on the sheets close to the tree, and with a hammer applies one or two sharp blows against a spike or large nail driven into the body of the tree, up from the ground as high as a man's head. Some have recommended to saw off a limb and strike against the end of the stub; but a spur of iron is preferable. Bore a half-inch hole one and a half inches into the side of the tree, and drive in an iron pin, say three or four inches long. The iron will not injure the tree, and the pin will last for many years—until the projecting part is grown over. If the body of the tree is struck with a beetle, axe-head, or a battering-ram, a serious bruise will be made; and more than this, a dull blow from a muffled mallet will not jar a tree sufficiently to make the curculios relinquish their hold. A sharp, jarring blow is essential. A blow or two with a hammer or an axe-head on an iron pin will bring down every curculio that may be on the tree, without injury to the bark of the tree. If trees are only five or six years old, a large nail, driven half way in, will subserve an excellent purpose. When two blows are given, they should follow each other in quick succession, as one blow may not always induce every curculio to quit his hold. But another blow, given as quickly as a person can strike with a hammer, would have the effect to bring down many more insects than a single blow. When fruit-trees are of numerous sizes—some small and others large—much judgment must be exercised in gauging the force of the blows. A light blow with a nail-hammer of ordinary size will be sufficient for many trees six to eight years old. A person of acute perception will soon be able to learn whether the blows given are too heavy or too light. In some instances, where trees are larger, it will be advisable to insert an iron pin, say three-fourths of an inch in diameter, into each large limb. The end of all such iron pins that enters the wood should be square, so that oft-repeated blows may not drive them deeper.

When sheets are spread on the smooth ground, every insect that drops can be seen readily; and if the number were large, the ends of the sheets can be gathered up quickly, and all the "Turks" be crushed in a moment. One person can manage such an apparatus as this with facility; and it will be found much more convenient, all things considered, than if the sheets were stretched across frames.

Catcher on Wheels.—Several Western pomologists have recommended the use of a large umbrella-like sheet, supported by a frame-work similar to the frame of a parasol, all of which is to be mounted on a kind of wheelbarrow (Fig. 92, on the opposite page), with a battering-beam in front. The vehicle is then run forcibly against the body of a tree, while the sheet is extended to catch the depredators as they fall.

The objections to such an apparatus are its cost, inconvenience in handling, and its inefficiency for giving a sharp, jarring blow to the tree. Besides this, when the sheet is spread, if supported two feet or more above the ground, a person can not approach the tree to jar it with a hammer. Every person will find that it will be most conven-

Fig. 92.

Curculio-catcher.

ient and economical, in every respect, to lay the sheets on
the smooth ground.   There will be many places where the
spread wheelbarrow-curculio-catcher can not be used without great inconvenience.   More than this, bunting the bod-

ies of trees, even with a cushioned mallet, or rammer, will injure the bark, unless the cushion is very thick, in which case the jar will not be sufficient to bring down the insects.

When to catch Curculios.—These depredators are found on fruit-trees all through the growing season. They seem to be illustrious philosophers. They will seldom deposit an egg in a cherry, plum, apple, or any other fruit, after the specimens have grown to a certain size, or have attained a certain age. If the egg is not deposited within a certain period after the blossom has fallen, or after the young fruit has formed, the growth will maintain the ascendency over the slow progress of the little larva in the fruit. In other words, the fruit will grow faster than the larva can eat. Hence the curculio will seldom deposit its eggs in fruit that is larger than the eggs of a quail. The young depredators lay their course always directly for the delicate central organs that run through the core of the apple or pear. Hence, if the putamen or stone of the plum, cherry, or peach, has begun to harden, the curculio seems to know it, and will not deposit eggs in such fruit. The eggs must be deposited before the putamen or shell of the pit has begun to harden. These suggestions warrant the direction to jar the trees daily for curculios before the young fruit is as large as peas of good size, until the last depredator has been captured, or until the putamen of the growing fruit is so hard that the young larva will not attempt to eat his way to the centre of the fruit. Let the spikes or stubs be inserted in the trees during the winter, when other duties are not urgent, and let the sheets be prepared long before they will be wanted. Then, by appropriating a little time daily to the work of extermination, the fruit can be saved at a small expense.

United Efforts for Extermination.—As the curculio has wings and can fly—no one knows how far or not how far— near neighbors should make a united effort to capture every invader, and to let not even *one* larva escape.   In some sections of country this has been done, and the result has been excellent crops of fine fruit.   When sheep and swine are not permitted to devour all the premature specimens that fall from the trees, such fruit should be picked up every day and destroyed, so as to prevent the increase of the curculio.   By united efforts, this invader could be effectually exterminated from the country in a few years.   They who would succeed in this work of extermination, must persevere long, as new crops of the insects often continue to come after the earlier ones are all destroyed.   The best time is early in the morning, when these insects are more torpid than at midday.   Once a day will commonly answer, except in seasons of extraordinary abundance, when a second examination should be made at sundown.   The work should not be intermitted a single day.   It is such intermissions that often cause failure.   Many have failed also by trying to shake the pest off the trees.   Nothing but a sharp jar will bring them down.

The Tent-caterpillar.—This pest of the apple-orchard is often alluded to as the most formidable enemy that pomologists have to encounter.   But this caterpillar may be exterminated with less trouble and labor than almost any other insect of the orchard.   The only remedy is to "catch 'em and kill 'em."   Unlike some other depredators, this one can be combated by destroying the moths, the eggs, or the larvæ, or full-grown worms.   When numerous, it has been known to strip whole orchards of their leaves, thus destroying the fruit-crop for the season, and sometimes proving fatal to the trees.   In many sections of the country where these caterpillars are not molested, we have often seen large orchards in the summer which looked as if a fire had been sweeping through the tree-tops.

The Tent-caterpillar is hatched from eggs deposited by a

11*

Fig. 93.

Moth of the American Tent-caterpillar
—male and female.

large miller, called the American Lackey-moth, of a yellowish brown color, represented by Fig. 93, which is often seen flying about among fruit-trees during the latter part of summer or in September, in our latitude. The American Lackey-moth, when fully developed, measures about one inch and a half from tip to tip of the expanded wings. It is usually of a pale brick color; but individuals are occasionally seen much darker, or of an ashy-brown color. Across the fore wings are two straight, oblique whitish lines. The antennæ are moderately pectinate, or feather-like, in the male, and very slightly so in the female. The hollow tongue, or sucker, through which insects of this order imbibe their nutriment, is wholly wanting in this species, as, indeed, it is generally in the particular group to which it belongs. Their short lives have but one object— the pairing of the sexes and the deposition of the eggs by the female for a future generation. The following experiment illustrates some of their habits: Three female moths were inclosed in a glass vessel. They were quiet during the day, but became very restless as night approached, showing that, like the moths in general, they are nocturnal in their habits. On the third day, a twig of an apple-tree was introduced into the vessel. The moths immediately ran up upon it, and put themselves in position for laying their eggs. This was accomplished in the following manner: Placing herself transversely upon the side of the twig, she curved her abdomen under the twig, and extended it up the opposite side as far as she could reach, and commenced depositing her eggs, one after another, gradually withdrawing the abdomen till she had laid a row of eggs across the under side of the twig. She then, in the same manner, deposited another row, parallel to and in contact with the first. Owing to their unnatural situation, or the absence of the opposite sex, or to some unknown cause, these moths in confinement succeeded in laying but two or three rows of eggs; while in a state of nature they lay from fifteen to twenty rows, containing in all an average of about two hundred and fifty eggs. They subsequently cover the eggs with a coating of brown varnish, which effectually protects them from the vicissitudes of the weather. Let every one be caught and killed, and thus destroy a large number of caterpillars. Let several bottles of sweetened water be placed near every tree, as directed (Fig. 51, p. 117). By this means a large number of caterpillars may be wiped out before they have had an existence.

This yellowish-brown miller, as already stated, deposits her eggs around small twigs, as shown at *c*, Fig. 95, p. 253, or by Fig. 94, herewith given, all placed on end, and covered over with a varnish, to shield them from rain and snow. Each bunch can readily be seen; and when taken off and burned, a whole nest will be destroyed at once, without resorting to soapsuds, coal-oil, or any other

Fig. 94.

Eggs of the Lackey-moth wound spirally around a twig.

application. This can be done at any time during winter or spring, before the tree leaves out. Should any be overlooked, they are easily destroyed about the time the buds commence to put forth, as the young caterpillars hatch out the first warm *spell* in the spring, and can always be found at the first crotch, or fork, below the eggs. In no case, however warm or protracted the autumn may be, do these eggs ever hatch till the following spring; so that the Tent-caterpillar, unlike many of our noxious insects, never has more than one brood in the season. At this stage of their existence they are confined to a very small compass, and can easily be destroyed. The best time to do this is early in the morning, or towards evening, as at these times they are all collected together, and quick work can be made with them. When out of reach, some use a pole, rough at the end. With this one can wind a whole web, worms and all, and thus exterminate the caterpillar.

We have frequently placed a knot of such eggs in a warm room during cold weather, where every egg has produced a minute caterpillar not one-eighth of an inch long. As soon as they are hatched, they begin to feed upon the tender leaves of the apple and some other trees, and increase in size and capacity for destruction with the growth of the foliage, destroying it as fast as it grows. In the early morning, while the dew is on the foliage, sprinkle fine air-slaked lime freely over the tree. The caterpillar will drop almost as soon as touched by the subtile dust, or perish while holding to the leaf, provided each one gets thoroughly dusted with the strong lime. But it is better to examine the outer twigs of trees several times for the nests. Then, it is better still to destroy the millers. In the spring, keep an eye on the trees for these depredators; and as soon as a nest is discovered, climb up and crush every one. Do not attempt to burn these pests, or to blow the nests to fragments with powder. Put on a leather mitten, and crush them long before they are large enough to injure the leaves. Do not permit a nest to be built in the orchard. Retire one hour earlier than usual, so as to be able to rise with the lark, and spend an hour crushing caterpillars before the sleepers have opened their eyes. This is the only remedy—" catch 'em and kill 'em." Large caterpillars will often crawl half a mile on a fence. Let such itinerant interlopers be crushed, as each one will change to a brown miller, like Fig. 93, p. 250, which will lay an untold number of eggs.

Most of our rural readers will recognize at a glance the perfect representation, given in the accompanying diagram, of a small nest of these· depredators of the apple - or-chard, two of which are full - grown, as shown at *a* and *b*. The nest is built by spinning numerous webs at the fork of two branches, from one to the other, leaving a hole for the entrance of the caterpillars on one side, as represented by the dark spot on the nest, between the caterpillars *a* and *b*. Many nests, when the. depreda-tors are not molest-ed, contain two to

Fig. 95.

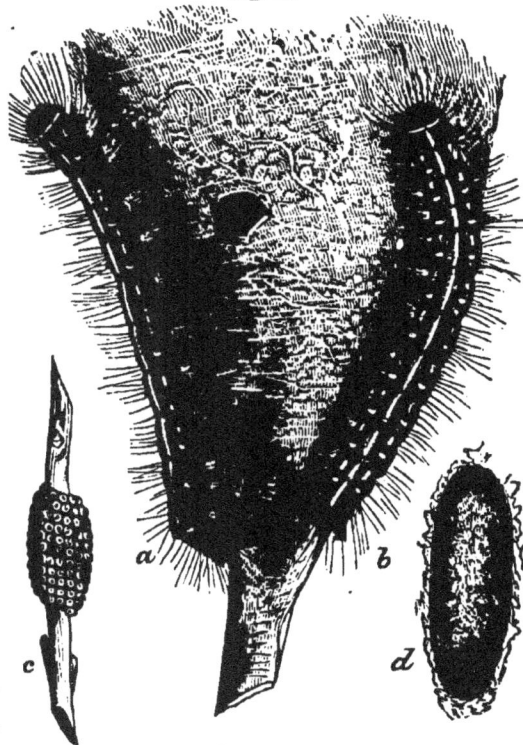

A nest of Tent-caterpillars.

four quarts of these repulsive invaders of the apple-orchard. If any of the larvæ—*a* and *b*—escape being crushed or de-stroyed in some manner, let them be hunted in the pupa state—as at *d*—or let their eggs, *c*, be destroyed. It is a singular fact that, before these eggs will hatch, there must have already been sufficient warm weather to expand the buds and develop small leaves. The young caterpillars are usually ready to devour the tender leaves as soon as the latter are of sufficient size for the depredators to get a bite. These larvæ grow rapidly when they can reach a liberal supply of young leaves. As soon as they fill themselves they return to the nest, and rest until their food is digested and a quantity of webbing material has been concocted, when they spin and weave up all their warp and woof, and then hasten away to feed again on the leaves. During cold nights and stormy days they collect in the nest, or where the nest is being built, and resist the storm as well as they can. They should never be permitted to build such a nest as is here represented, as their ravages are exceedingly injurious to the growing tree, especially to its fruitfulness. If unmolested, these caterpillars will soon become large and fat, and most of them will leave the nest and enter the pupa state, *d*, and some will crawl on fences a long distance.

As soon as their career in the larva state is run, each one spins a covering around its body, in some secluded place where storms can not beat upon it, as shown at *d*, which is a cocoon containing the chrysalis, or parent insect, in the pupa state. During the months of July and August, these cocoons may be seen beneath pieces of bark, on fences, attached to gate-posts or out-buildings. If all such cocoons be destroyed, no larvæ (caterpillars and moths) would be seen the next season. If the cocoons—*d*, in the illustration—are not disturbed, the parent insect, in the form of a large miller, as already stated, will soon emerge, and commence the task of its little life—laying the eggs, *c*.

In some sections of country, during the middle and latter part of summer, some orchards and vineyards seem to be alive with moths and millers of the kind alluded to, a large proportion of which may be exterminated with little effort and expense, by trapping the depredators in small bottles partly filled with very sweet water, to which a little vinegar has been added.

When the Tent-caterpillars leave their nests for the pur-pose of going to feed on the leaves, they travel upon the upper side of the branches; and each one leaves a thread of silk behind it, which probably serves as a clue to direct it back to the nest. The silken trails thus formed are at first scarcely noticeable; but they become very obvious after a branch has been travelled upon for a considerable time. Thus the caterpillar not only lives in a silken house, but covers its roads with a silken carpet. Like other larvæ, they shed their skins four times before arriving at maturity. When fully grown, they are about two inches in length, and frequently two inches and a half long, fat and round. The "American Entomologist," for which Fig. 95, p. 253, was originally engraved, states that the eyes of this cater-pillar have the appearance of very minute black points, being ten in number, five on each side of the head. Their position is best seen by holding the cast-off skin of a cater-pillar towards the light and examining it through a magni-fying-glass. Without claiming mathematical exactness, it may be stated that four of them are situated in a curved line, forming half a circle, of which circle the fifth occupies the centre. Owing to the extreme minuteness of the eyes of caterpillars in general, they were formerly overlooked, and these insects were supposed to be blind. That they possess the sense of seeing, however, and that, too, at a con-siderable distance, seems to be proved by the following ex-periments. If a nest of these caterpillars be taken from a tree and placed upon the ground several feet from it, they will return to the tree in a direct line. In another experi-ment a handful of caterpillars was placed in some tall grass between two trees, but nearer to one than the other. They first crept up the stems of the grass, as if for the purpose of taking an observation, and then took up their march for the nearest tree.

The leaves of the apple-tree constitute the food of by far the greater number of this kind of caterpillar. Nests are, however, occasionally seen on the other common fruit-trees —the peach, pear, plum, and cherry, particularly the wild cherry. When deprived of other food, they will also eat the leaves of the rose-bush.

The active period of this caterpillar, that is, the time from their hatching to their changing into chrysalids, is from five to six weeks; and when we consider their voracious appetites, and that there are about two hundred and fifty individuals in each nest, we can easily form an idea of the extent of their ravages. Where there happen to be several nests on one tree, or where the tree itself is small, they often strip it of every vestige of foliage; and in neglected localities whole orchards are sometimes seen as bare of foliage on the first of June as in mid-winter. When about to construct its cocoon, the insect attaches itself by its hindmost feet, so as to leave the anterior part of its body free for motion; then extending its body, it draws some disconnected lines across from one side of the angle to the other, to serve as outlines or stays. Then, working down nearer home, it draws its lines more densely, so that near its body they constitute a pretty close texture, like a piece of loosely-woven cloth, through which, however, the insect can be seen. When the web is finished, the insect emits a yellow fluid, with which it besmears the inside of the cocoon, and thus effectually conceals itself from view.

The Apple-tree Borer.—An untold number of choice apple-trees have been destroyed every year by Apple-tree Borers, when the proprietors have wondered what could have killed their trees? And hundreds of young apple-trees all over the country may be found, at the present day, that have been infested with borers for one or more years, which have injured the growing trees so seriously that they will never recover. In our visits through the country, we frequently see orchards that are half ruined by borers; and the proprietors had not even thought that there was a borer to be found. We have frequently been asked if we could not suggest some good reason for the singular appearance of certain trees; whereupon, to give ocular demonstration that borers were at work, we have applied the point of a jack-knife, dislodging, in some instances, eight or nine borers from one tree. Any intelligent observer may, by close attention, soon learn how to discover, almost at first glance, the place where a newly-hatched borer lies concealed. Usually, the Apple-tree Borer is found at the collar of the tree, close to the ground; and they are some-times detected on the body and large branches of a tree. A small drop of brown fluid, resembling tobacco-juice in color, usually reveals its presence; for at that early stage of its development the well-known sawdust-like excretions characteristic of the full grown larvæ must not be looked for.

The Apple-tree Borer is the larva or grub of a beautiful bug or beetle, which deposits its eggs at or near the surface of the ground, on the trunk of the tree, during the months of May and June; and sometimes, entomologists assure us, the eggs are oviposited in July and August by the parent bugs (Fig. 100, p. 261), and in a week or two little plump borers are hatched, which immediately begin to bore through the tender bark.

The next spring, as soon as hard freezing is over, he goes to work vigorously, still feeding on the liber, or inner bark. By fall he will have attained a length of one-half to three-fourths of an inch. Some entomologists have stated that the borer does not emerge from the tree until the third summer. But Professor Fitch, entomologist of the State of New York, says that

"The beetle comes abroad in June, and drops its eggs under the loose scales of the bark, low down near the surface of the earth. The worm which hatches therefrom eats inward through the bark, till it comes to the wood. It there remains, feeding upon the soft outer layers of the wood, and thus excavates a shallow round cavity under the bark, the size of a half-dollar; though where two, three, or more worms are lodged in the same tree, as they always preserve a narrow partition between their cells, one never gnawing into that of another, these cells, by crowding upon one another, become of an irregular form, and almost girdle the tree. The cell is always filled with worm-dust, crowded and compacted together, some of which becomes crowded out through a crack in the bark, or a hole made by the worm. And it is by seeing this sawdust-like powder protruding out of the bark that we detect the presence of these borers in the tree. The worm continues to feed and enlarge its cell under the bark for about twelve months, until it has become half-grown, and is from a half to three-fourths of an inch in length. Its jaws have now acquired sufficient strength for it to attack the solid heart-wood of the tree, and it accordingly bores a cylindrical hole from the upper part of its cell, upward in the solid wood, to a length of three or four inches or more, this hole inclining inward towards the centre of the tree, and then curving outward till its upper end comes again to the bark. It then stuffs the upper end of this passage with fine chips or worm-dust, and its lower end with short fibres of wood, arranged like curled locks of hair, thus forming an elastic bed on which to repose during its pupa state. These operations being completed, it throws off its larva skin and becomes a pupa, usually at the close of the second summer, or about fifteen months after it is hatched from the egg. In this state it lies through the winter, and changes to its perfect form in the following spring; but often continues to lie dormant several weeks after its final change, until the season becomes sufficiently warm for it to come abroad. Awaking then into life and activity, it crawls upward, loosening and pulling down the chips and dust that close the upper end of its burrow, till it reaches the bark. Through this it cuts with its jaws a remarkably smooth, round hole, of the exact size requisite to enable it to crawl out of the tree. The sexes then pair, and the female deposits another crop of eggs."

Operations of the Apple-tree Borer illustrated.—Fig. 96, on the opposite page, represents a section of a young apple-tree which has been bored by several of these depredators. The dark spots represent the holes which his sharp teeth have gouged out in the solid wood. These little puny worms destroy whole orchards. After Apple-tree Borers once get beyond the reach of the knife, it is exceedingly difficult to dislodge the depredators and save the trees.

Efficacious Remedies.—"Catch 'em and kill 'em" is the only remedy that we have ever known to be really efficient.

The beetle that de-
posits the eggs is a
wise philosopher, as
to selecting a con-
genial place for ovi-
positing the eggs
for the grubs. The
bark of a young ap-
ple-tree close to the
ground is exceed-
ingly tender at the
collar; consequent-
ly the young larva
can readily cut his
way through the
bark to the wood.

Fig. 96.

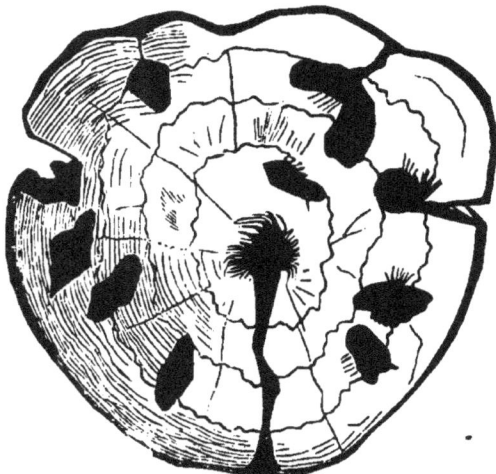

A transverse section of the stem of an apple-tree
badly bored by the Apple-tree Borer.

Every tree should be examined carefully at least every
month, during the growing season. As a "workman is
known by his chips," so is the borer traced by the dust or
cuttings which he throws out behind him. Clear away
the grass and weeds from the trunk and exposed roots of
the tree, but do it carefully, as the cuttings or chips eject-
ed by the young grub will often be found standing up
quite erect from his entrance, being glued together in some
manner by the industrious borer. By scraping off the bark
carefully, you will find the grub in a snug little bed upon,
or slightly sunk into, the bark. It will be only a small job
to destroy every invader the first season, as they are still in
the bark, where birds often take out every one. Woodpeckers frequently peck holes through the bark and take
them out. One or two months after the larvæ have commenced boring, a person can see exactly where every one is
located, and can kill them with the point of a penknife.
But, after they have been boring a year or more, the true

way is to employ a small gouge and mallet, and cut grooves up and down the tree, until every borer is removed. Then fill every recess with grafting-wax, and bind it around with cloth. We might record a dozen other remedies which will not be effectual. But we know cold steel will never fail to do the job effectually.

A Preventive of the Borer.—In case a person is certain no eggs have been deposited in his young trees, it will not be difficult to keep beetles from depositing their eggs in the bark by scraping away the earth an inch below the collar of the tree, then wrapping two or three thicknesses of heavy brown paper around the body of the tree for two or three feet high, tying it in three places with elastic wrapping-thread, and smearing the paper with a heavy coat of coal-tar. (See page 166.) No beetle or bug will deposit eggs in a tree thus protected. It is said that the bugs are very shy, and work only in the night. But we have often met with them in the day-time, and have easily caught them.

Fig. 97 represents the parent bug, or beetle of the Flat-headed Borer (*Chrysobothris buprestis*), and Fig. 98 is the borer of life size. Although the Flat-headed Borer evinces a manifest partiality for the various sub-varieties of the *Pyrus malus* trees and *Pyrus baccata*, as well as for our own indigenous crabs, it must not be imagined that it disdains other food. We have found these borers preying upon the pear, though seldom; and occasionally upon the Mazzard and other cherry-trees. The same means are used to combat both; although of course allowance must be made for the peculiarities in the habits and modes of life of each.

Fig. 97.   Fig. 98.

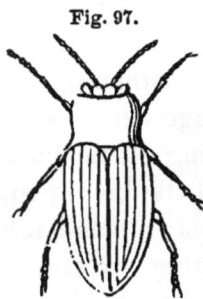

Parent bug of the Flat-headed Apple-tree Borer.   Flat-headed Apple-tree Borer.

The Round-headed Borer (*Saperda bivittata*), Fig. 99, is often found in any part of the body of an apple-tree, although the parent-bug (Fig. 100) will seek the tender bark at the collar of the tree first, as a suitable place for depositing its eggs. Some- times both species dwell togeth- er in the same orchard *and on the same tree ;* often, however,

Fig. 99.

Fig. 100.

Round-headed
Apple - tree
Borer.

Parent bug of the
Round - headed
Borer.

the Round-headed Borer will be found mainly to infest a cer- tain orchard, while another orchard, not a quarter of a mile off, will be exclusively attacked by the Flat-headed Borer.

It is interesting to dwell on the habits of such depreda- tors. But we must cut every thing short for the sake of giving practical directions for preventing the ravages of these pests. Red-headed Woodpeckers will destroy every borer that can be found. Hence, let such birds be encouraged to frequent apple-orchards. Then let the bug-traps be kept in running order, to destroy the beetles, the moth-millers, and the flies, when they are about to deposit their eggs or spin cocoons. After borers have really bored deeply into a tree, there is no other prac- ticable way to dispose of them than to cut them out with a joiner's gouge and mallet, and cover the wounds well with grafting-wax.

Fig. 101.

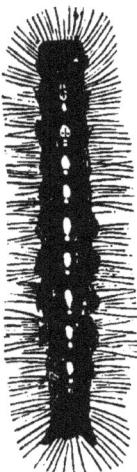

Forest Tent-
caterpillar.

The Forest-tent Caterpillar (*Clisiocampa syl- vatica*).—This caterpillar (Fig. 101) is frequent- ly confounded with the "American Tent-cater- pillar " (*Clisiocampa Americana*), Fig. 95, p. 253. Although its ravages are usually confined to the forest and groves, the depredators often appear in large numbers in apple-orchards, where they do great damage if permitted to flourish unmolested. Professor Fitch says that

"This caterpillar, as seen after it has forsaken its nest and is wandering about, is 1¼ inches long and 0.20 thick. It is cylindrical and of a pale blue color, tinged low down on each side with greenish gray, and is everywhere sprinkled over with black points and dots. Along its back is a row of ten or eleven oval or diamond-shaped white spots, which are similarly sprinkled with black points and dots, and are placed one on the fore part of each segment. Behind each of these spots is a much smaller white spot, occupying the middle of each segment. The intervening space is black, which color also forms a border surrounding each of the spots, and on each side is an elevated black dot, from which arises usually four long black hairs. The hind part of each segment is occupied by three crinkled, and more or less interrupted pale orange-yellow lines, which are edged with black. And on each side is a continuous and somewhat broader stripe of the same yellow color, similarly edged on each of its sides with black. Lower down upon each side is a paler yellow or cream-colored stripe, the edges of which are more jagged and irregular than those of the one above it, and this stripe also is bordered with black, broadly and unevenly on its upper side, and very narrowly on its lower side. The back is clothed with numerous fine fox-colored hairs, and low down on each side are numerous coarser whitish ones. On the under side is a large oval black spot on each segment, except the anterior ones. The legs and prolegs are black, and clothed with short whitish hairs. The head is of a dark, bluish color, freckled with numerous black dots, and clothed with short blackish and fox-colored hairs. The second segment or neck is edged anteriorly with cream white, which color is more broad upon the sides. The third and fourth segments have each a large black spot on each side. The instant it is immersed in spirits the blue color of this caterpillar vanishes, and it becomes black."

The accompanying illustration (Fig. 102) will give the reader a correct idea of some of the stages of existence through which this insect passes. At *a*, the egg-mass is represented as it is found attached to small twigs of apple - trees. These eggs may be readily distinguished from the eggs of the common American Tent-caterpillar, by the uniformly cylindrical diameter, and by being

Fig. 102.

a, the eggs of the Forest Tent-caterpillar; *b*, the parent miller that lays the eggs; *c*, transverse section of one of the eggs; *d*, magnified.

docked-off squarely at each end. In each of these masses there are about four hundred eggs. Hence, as every egg will produce a caterpillar, as certainly as a kernel of Indian corn will produce a stalk, by cutting off every cluster of eggs at any time before they hatch, about four hundred hungry caterpillars will be destroyed at one stroke.

The eggs are deposited, in some localities, during the latter part of June. The embryo develops during the hot summer weather; and the yet unborn larva is fully formed by the time winter comes on. The caterpillars hatch with the first warm weather in spring—generally from the middle to the last of March, and in April — and though the buds of their food-plant may not have opened at the time, and though it may freeze severely afterwards, yet these little creatures are wonderfully hardy, and can fast for three whole weeks, if need be, and withstand any amount of inclement weather. The very moment these little larvæ are born, they commence spinning a web wherever they go. At this time they are black, with pale hairs, and are always found either huddled together or travelling in file along the silken paths which they form when in search of food. In about two weeks from the time they commence feeding, they go through their first moult, having first grown paler, or of a light yellowish-brown, with the extremities rather darker than the middle of the body, with the little warts which give rise to the hairs quite distinct, and a conspicuous dark interrupted line each side of the back. After the first moult, they are characterized principally by two pale yellowish subdorsal lines. After the second moult, which takes place in about a week from the first, the characteristic pale spots on the back appear, the upper pale line becomes yellow, the lower one white, and the space between them bluish. Very soon they undergo a third moult, after which the colors all become more distinct and fresh; the head and anal plate have a soft, bluish, velvety appearance, and the hairs seem more dense. After undergoing a fourth moult without material change in appearance, they acquire their full growth in about six weeks from the time of first feeding. At this stage of development, the larva appears fully grown, and may be seen wandering singly over different trees, along roads, or on the tops of fences, in search of a suitable place to form its cocoon. It usually contents itself with folding a leaf or drawing several together for this purpose, though it frequently spins up under fence boards and in other sheltered situations. The cocoon is very much like that of the common American Tent-caterpillar, being formed of a loose exterior covering of white silk, with the hairs of the larva interwoven, and by a more compact oval inner pod that is made stiff by the meshes being filled with a thin yellowish paste from the mouth of the larva, which paste, when dried, gives the cocoon the appearance of being dusted with powdered sulphur. Three days after the cocoon is completed, the caterpillar casts its skin for the last time, and becomes a chrysalis of a reddish-brown color, slightly dusted with a pale powder, and densely clothed with short, pale-yellow hairs, which at the blunt and rounded extremity are somewhat larger and darker. In a couple of weeks more, or during the forepart of June, the moths commence to issue, and fly about at night. This moth bears a considerable resemblance to that of the American Tent-caterpillar (Fig. 93, p. 250).

From the very moment it is born till after the fourth or last moult, this caterpillar spins a web, and lives more or less in company; but from the fact that this web is always attached close to the branches and trunks of the trees infested, it is often overlooked; and several writers have erroneously declared that it does not spin. At each successive moult, all the individuals of a batch collect and huddle together upon a common web for two or three days, and during these periods—though more active than most other caterpillars in this so-called sickness—they are quite sluggish. During the last or fourth moult they frequently come low down on the trunk of the tree, and, as in the case of the gregarious larvæ of the Hand-maid Moth (*Datana ministra*), which often denude our black-walnut trees, they unwittingly court destruction by collecting in such masses within man's reach. From the time they are born till after the third moult, these worms will drop and suspend themselves mid-air, if the branch upon which they are feeding be suddenly jarred. Therefore, when they have been allowed to multiply in an orchard, this habit will suggest various modes of destroying them. They can often be slaughtered *en masse* when collected on the trunks during the last moulting period. They will more generally be found on the leeward side of the tree, if the wind has been blowing in one direction for a few days. The cocoons may also be searched for, and many of the moths caught by attracting them towards the light. But pre-eminently the most effective artificial mode of preventing this insect's injuries is, to search for and destroy the egg-masses in the winter-time, when the trees are leafless. This course is the more efficient because it is more easily pursued, and nips the evil in the bud. Tarred bandages, or any of the many remedies used to prevent the female Canker-worm (p. 233) from ascending trees, can only be useful with the Forest Tent-caterpillar, when it is intended to temporarily protect an uninfested tree from the straggling worms which may travel from surrounding trees.

The Codling-worm Moth.—Doubtless most persons who are in the habit of eating crude apples have repeatedly noticed the little whitish worm which is so often found burrowing at the core of the fruit, and filling it with its dis-

12

gusting excrement.   But probably not one fruit-grower out
of a hundred has ever seen the little moth which is produced from this worm, and which, in its turn, gives birth to
a fresh generation of such worms.   This moth is variously
known as the Apple-worm Moth, or the Codling-worm Moth.
The Apple-worm Moth makes its first appearance about the
first of May to the forepart of June, and a little earlier or
later, according to the season or the latitude.   Usually, at
the time it appears, the young apples are already set, and
are beginning to be about as large as a hazel-nut.   After
coupling in the usual manner, the female moth then proceeds to deposit a single egg in the blossom end of the fruit,
flying from fruit to fruit, until her stock of eggs (amounting to two or three hundred) is exhausted.   Soon after accomplishing this process, she dies of old age and exhaustion.   In a very few cases, the egg is deposited in the cavity at the stem-end of the fruit, or simply glued on to the
smooth surface of the fruit.   In a short time afterwards,
the egg, no matter where it is located, hatches out, and the
young larva forthwith proceeds to burrow into the flesh of
the apple, feeding as he goes, making his head-quarters in
the core.   In three or four weeks' time, it is full-grown;
and shortly before this, the infested apple generally falls to
the ground.   The larva then crawls out of the fruit through
a large hole in the cheek, *l*, which it has bored several days
beforehand for that express purpose, and usually makes for
the trunk of the tree, up which it climbs, and spins around
itself a silken cocoon of a dirty white color, in any convenient crevice it can find, the crotch of a tree being a favorite spot.   Here it transforms into the pupa state, and
towards the latter part of July or the forepart of August
it bursts forth in the moth state.   It also appears in the
former part of the growing season, and it has been noticed
that a larva will occasionally spin its cocoon on the under-

surface of some board lying flat on the ground, instead of climbing the tree in the usual manner.

Fig. 103, which was originally prepared for the "Culti-vator," will furnish a large amount of information concern-ing this depredator of our apple-orchards in various stages.

Fig. 103.

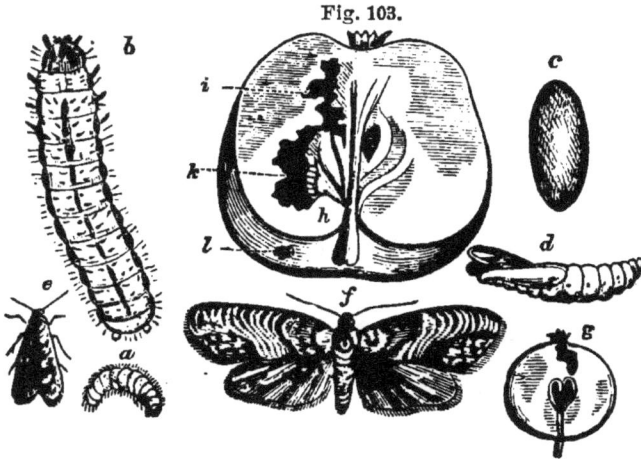

The Codling Moth, or Apple-worm.

At *a*, the larva of natural size is shown. The same is rep-resented at *b*, highly magnified. At *c* a cocoon is shown, and *d* represents the pupa within the cocoon. At *e* a per-fect insect is shown at rest. At *f* the same miller, or fly, is represented with its wings spread. At *g* a small apple is shown, with the young larva just entering the fruit. In many instances, when apple-trees are growing where the soil is rich in apple-producing material, the expansion of the fruit will often exceed the ravages of the larva that is eat-ing its passage to the core. If they can not reach the core, the fruit, in most instances, continues to grow, and the lar-va dies. We frequently see apples that are imperfect and much depressed on one side of the basin, the cause of which is usually attributable to the working of a larva when the fruit was very small. As the apple grew faster than the larva could eat, the invader perished, and the apple was injured but little.

Remedies for the Codling-worm Moth.—If the millers can be caught, of course no eggs will be deposited from which larvæ can be hatched. Hence, let the bottles of sweetened water and vinegar (Fig. 51, p. 117) be employed. A large proportion of these flies may be trapped in bottles and by fires, or by the lamp-trap (Fig. 104). After the larva, *k*, reaches the core of an apple, the fruit will usually drop from the tree before it has ripened. In a few days after this, the larva bores out of the apple and winds itself up in a cocoon, *c*, beneath shreds of bark, as already stated.

Fig. 104.

A nocturnal insect-catcher.

In August or September, or the next spring, these insects reappear, in time to deposit their eggs in the calyx of the young fruit. If little birds are numerous among apple-trees, they will devour almost every larva before they can produce a cocoon or find a refuge where they will be safe from birds. These larvæ are fat and tender worms, and all kinds of birds devour them with avidity. Hence the advantage of encouraging the propagation of these little songsters—like the yellow-bird, wren, robin, "little chippie," and many others that choose the fruit-orchard for their abode. Where little birds have been exterminated, apples are apt to be wormy. Some pomologists twist a rope of hay or straw, say half as large as a man's wrist, and wind it several times around the body of an apple-tree, as shown by Fig. 92, p. 247. Beneath this band the larvæ will take refuge and spin their cocoons, *c*. The band should be removed soon after the cocoons are formed, and the cocoons destroyed. In some instances, at the West, these depredators are so numerous that old cloths are laid in the crotches of the branches for a few weeks, then they are passed through a clothes-wringer to crush the cocoons, after which the cloths are returned to trap other insects. If swine and sheep are permitted to range about the trees, they will usually destroy the fallen fruit with the larvæ. When swine are not permitted to have access to the orchard, the fallen fruit should all be picked up and crushed, to destroy the larvæ. Hence we see that an incessant warfare must be maintained if the fruit is saved. These depredators must be caught and killed, or they will destroy the fruit. The habits of this insect are now so thoroughly understood, that it can be combated successfully in every stage of its existence. By keeping the bottles containing sweetened water (Fig. 51, p. 117), and the pan half-filled with thin molasses, with a lighted lamp near it (Fig. 104, p. 268), in the orchard every night, in good order, almost every insect will be trapped in a few days.

Apple-twig Borers.—In many sections of country, Apple-twig Borers injure valuable trees by boring small twigs, which often are so nearly cut in two that the wind breaks them off. The only remedy thus far discovered is to " catch 'em and kill 'em." A constant watch must be kept over every fruit-tree for insect enemies. Then, if any thing is discovered that resembles the work of the Apple-twig Borer, let the branches be examined with care, and the depredators destroyed. Cold steel is a certain remedy for this pest of the apple-orchard. As these borers originate from a parent insect that can be trapped in a bottle, the bottle-traps will be found the most satisfactory way of destroying this kind of depredators.

Bark Lice (*Aphis mali*).—In many orchards, bark lice are scarcely known; while in others, where trees are young and thrifty, bark lice cover almost every twig.   Dr. Fitch, entomologist of the State of New York, writes of the Apple-bark Louse (*Aspidiotus conchiformis*) that

"The Apple-bark Louse makes its appearance as a little brown scale, one-eighth of an inch long, the shape of an oyster-shell, fixed to the smooth bark, resembling a little blister.   This scale is the dried remains of the body of the female, covering and protecting her eggs, from a dozen to a hundred of which lie in the cavity under each scale.   These eggs hatch the latter part of May, and the young lice diffuse themselves over the bark, appearing as minute white atoms, almost invisible to the eye.   They puncture the bark, and suck the sap from it.   The females soon fix themselves and become stationary.   They die, and become overspread with a substance resembling fine blue mould, which wearing off, the little oyster-shaped scale again appears in July.   They sometimes become so multiplied that the bark of the trunk and limbs is everywhere covered and crowded with them ; and if the tree is weakened by borers, fire blight, or other disease, these bark lice, thus multiplying, kill it."

Numerous remedies have been suggested for the Bark Louse, among which are the following: The bodies of the trees and large branches are scraped, if covered with rough bark, and afterwards smeared with thin soft-soap applied with a swab.   Some have used equal parts of pine-tar and linseed-oil boiled together, and applied when warm and thin. We have employed equal parts of pine-pitch and linseed-oil, or tallow mingled together by heating, which was applied with a swab.   If there is much rough bark on the trees, it must all be removed ; and if the surface is shaved smoothly before the material is applied, all the better.   We never find bark lice attached to thick and tough bark.   They must have tender and thrifty bark, which they can puncture, in order to extract the sap.   If the young and tender bark be thoroughly soaped, and a smooth surface formed, without being protected by liquid pitch, all the soap will be removed during a heavy rain, and the bark lice will be provided with just such quarters as they desire; but the covering of pitch and linseed-oil will form a varnish which will thoroughly exclude every thing that is ever found adhering to the bark of apple-trees.   No borer will ever attempt to deposit an egg in the bark that is varnished.   As this varnish will crack when the growing trees expand, let more varnish be applied early every spring, and bark lice will not touch a tree. Tobacco-water, soda-wash, and quassia only prepare the bark for the occupancy of these depredators, rendering the surface soft and tender. Two dimes' worth of pitch and tallow will be of more value in excluding lice than a dollar's worth of soap or potash-water.

Some pomologists have stated that they have employed a decoction of tobacco, with excellent results, in the following manner: Procure a few pounds of leaf-tobacco, and boil it to an impalpable pulp, which is afterwards mingled with cold soft-soap, and worked until the mass is of the consistence of thick paint. The rough bark is then scraped off the trees, after which a heavy coat of this daubing is applied both to bodies and large branches. This material is excellent for the purpose intended, so long as only a thin layer of it will remain on the tree; yet two or three heavy and protracted storms will wash nearly all of it to the ground. But the liquid grafting-wax will adhere for several years. The expense of preparing and applying it will not exceed the expense of the tobacco and soap.

Fig. 105.

The Oyster-shell Bark Louse.—Figure 105 represents a small part of a branch covered with the eggs of the Oyster-shell Bark Louse. The only true remedy—we repeat—is to scrape them off and coat the smooth bark with liquid grafting-wax. Lice are wonderfully destructive to both the growth and health of trees; and the rapidity with which they increase when unmolested is astonishing. They are the color of the bark itself, and in shape and size like a flax-seed. In frequent cases they actually kill the branch they settle upon; and in young trees often kill them

Oyster-shell
Bark Lice.

altogether. The Oyster-shell Bark Lice are to be dreaded more than any other, as they do greater injury to young trees. This pest is said to be migratory in its habits, often attacking the thriftiest as well as the weakest trees, and its progress should be arrested before it ravages the whole orchard. Bark lice are devoured by millions by wrens, chickadees, and other similar birds; and "lady-bugs" also destroy large numbers of them.

The Scale Insect (*Homoptera*) is one of the many enemies of the apple, belonging to a family that contains more anomalous forms than any other. All this family are supplied with a suctorial mouth, arising so far back on the under side of the head as apparently to come from the breast in some species. The present insect is included in the genus *Coccus*, and has for its near relations some that have been useful to man from the time of the ancients, producing valuable dyes, the cochineal being one of them; and it is calculated that in one pound of this dye there are 70,000 of these little insects. When first hatched from the egg, it possesses considerable ambulatory powers, and can crawl all over a tree and select a situation. It then inserts its *rostrum* into the tender bark and draws the sap; and such a constant drain, by the countless numbers found upon a tree, must be very injurious. The insect remains in this position until death, in the female, undergoing its transformations, which, instead of producing a higher state of development, as in most other forms, has a contrary effect, it becoming, in fact, a mere inert, fleshy mass, in some allied species losing even the rudiments of limbs and all appearance of articulation. The male, on the contrary, which is much smaller, in casting off his pupa skin, obtains pretty large wings, and well-developed limbs, armed with a single claw, and his mouth becomes obsolete; he then sallies forth in search of his partner, of which he sees nothing but the pupa envelope. The female afterwards becomes distended with eggs. She then gradually dries up, leaving the shell of her body for a covering to the newly-hatched young, of which there are two broods in a year.

Harris recommends a preventive: To two parts of soft-soap add eight of water, and mix as much lime with it as will make a stiff whitewash, and apply with a brush to the trunk and branches of the infected trees, in the month of June, when the young insects are newly hatched.

Root Lice, how to Exterminate.—In many sections of country, roots of apple-trees, and some roots of other fruit-trees, are infested with immense numbers of root lice. The most effective remedy is to apply liberal quantities of ashes, lime, or barn-yard manure, and dig them into the soil round about the trees, as far as the branches extend. Dr. Warden recommends boiling cheap tobacco, or the stems, in water; after the strength has been extracted from the tobacco, skim out the stems and leaves, and to each bucket-ful of the water add, say, one quart of soap. When the soap becomes mixed, and the decoction is sufficiently cool-ed, it is fit for use. A good plan is to take a barrel to con-tain the mixture to a central part of the ground, where the trees are to be planted, and when one lot of trees is taken out for planting, another lot may be put in. In this way, with but little loss of time, the trees will be immersed long enough to kill all the lice there may be on the roots. Tree-roots once free from lice may be kept so by smearing the trunks above ground, each year, with boiling-hot soap. This should be done in the month of June, when it will answer the double purpose of keeping away the lice, and preventing the Apple-tree Borers from depositing their eggs. The soap put upon the trunks of trees, while it is hot, strikes into the bark, and is not soon washed out. Make a liberal use of the soap at the base of the trunks, as it is at these parts that the *seperda* prefers to lay its eggs; and just below the surface it is that root lice assemble before going upon the roots for the winter. In some instances, where the soil is unusually light and porous, two or three wagon-loads of clay spread around a tree, and worked into the soil, will operate as a satisfactory preventive of lice.

After root lice have deprived a root of its juices, they shift to others which will afford the needed supply of food. Thus one root after another yields up its nutriment to these pests, until the tree is either killed, or so weakened that it is attacked in all its parts by borers. In the summer numbers of them may always be found in the tops of the trees, under the partial cover of new wood and bark growth, where it is forming over fresh wounds made in pruning. When considerable numbers of lice assemble at these points, they cause numerous warty excrescences on the new-made bark, similar to the parts punctured by them below ground. On summer-grafted or newly-budded trees, they are a great nuisance. They soon find the wounds made by inserting the grafts or buds, which they enter, and, if not prevented, they so deplete the parts that the stock will not unite with the graft or bud. If a heavy mulch of straw or hay is placed close around the trees early in the fall, vast numbers of lice will come together on the tree at the surface of the ground. On single trees so treated, we have, in the month of October, seen as many as half a pint of these insects. It has been suggested that advantage might be taken of their coming together under the mulch by pouring boiling water over them. But if they are allowed to remain long at this point, they will kill the tree by severing the connection between the sap-vessels in the top and roots. On large roots the injuries done by root lice do not, at first, become apparent; generally not until after the lice have left, and then only by the dead roots which they leave behind. Small roots, when punctured by them, become knotty and greatly deformed; but large roots, at the collar of the tree, may be sucked dry by them without showing knots. On this account, the real cause of the death of the tree is often overlooked. Indeed we have known persons possessing much entomological information inspect trees killed or damaged without for once suspecting that root lice had any agency in producing the results. Probably there are but few persons who are aware to what extent the Woolly Root Louse injures apple-trees, or how readily they discriminate between roots of trees which are healthy and those that are wounded.

The Woolly Aphis.—This scourge of young apple-orchards (Fig. 106) is sometimes alluded to as the "American Blight." But entomologists assure us that it is a species of Aphis, or Plant Louse. We frequently see it on the branches of young trees covered with fine white downy hairs. In some instances, they are so numerous as to destroy a streak of the bark on a limb the entire length. They are voracious feeders. They suck out the juice of the tender branches with surprising rapidity. We have seen branches of trees with a dark-colored streak for several feet in length, where these depredators had been destroyed, which appeared as if a red-hot iron had been drawn along the bark.

Fig. 106.

Woolly Apple Louse.

The only true mode of extermination is to apply whale-oil soap with a swab, or strong lime-wash, rubbing the affected parts until every vestige of the depredators is wiped out. Then let the streak of injured bark be covered with liquid grafting-wax, or pitch and tallow.

Aphis, or Aphides.—We frequently see untold numbers of the aphis (represented by Fig. 107, on the following page), which is about the usual size of these depredators, on the tender branches, twigs, and leaves of apple-trees in the spring and summer. They puncture the surface and

Fig. 107.

The aphis.

suck out the sap, and almost stop the growth of the entire tree. In Fig. 108, c and D represent aphides magnified. These depredators are astonishingly prolific. Professor Reaumur showed, during his investigations, according to his computation, that

Fig. 108.

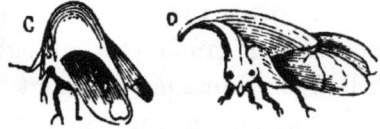

The aphis magnified.

one single aphis may produce, in five generations, if none were destroyed, more than six thousand millions of descendants! But, allowing that two-thirds of them are destroyed before they have attained sufficient age to propagate their young, two thousand millions are left to prey on apple-trees, leaves, and tender spray. They frequently cover the entire surface of leaves and twigs, and they injure the leaves to such an extent that they cease to grow, and the remaining leaves roll up and die. Other insects, and ants in particular, destroy large numbers of the aphis. When they are numerous, let the branches be sprinkled, by means of a syringe (Fig. 109), with strong soap-suds several times a day, in cloudy weather. In some instances, let the branches be washed and rubbed with a cloth. Leaves should not be sprinkled when the sun shines, in a hot day, as they are liable to be scalded by the liquid and heat of the sun. Early in the evening is the better time to apply the suds. If trees are small, many of the branches may be bent down and held for a moment in a tub of strong suds, which will effectually destroy the lice.

Fig. 109.

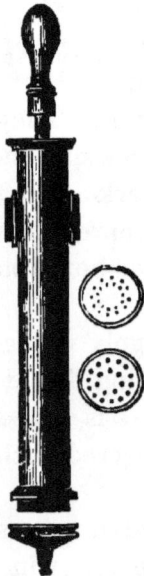

A hand syringe.

Fig. 110.

Douglas's aquarius.

Fig. 110 represents an *aquarius,* or hand water-engine, invented by W. B. Douglas, of Middletown, Connecticut, which we know, from experience, to be an excellent device for sprinkling trees with soap-suds, or for watering flower - beds, and even for washing the outside of windows. A small lad can sprinkle soap-suds all over a tree twenty feet high with such an aquarius. The illustration scarcely requires an explanation. The nose is provided with a *rose-spout,* full of numerous small holes, for the purpose of scattering all the liquid into spray. Such an aquarius may be obtained at most hardware stores.

Apple-tree Worms.—One of the most difficult apple-tree worms to combat is the small green sixteen-legged larva, nearly half an inch long, and with a broad, dark-brown stripe on each side, extending the whole length of its back. These depredators are found to do considerable damage to the apple-tree. They are said to belong to a new and hitherto undescribed species. The mode in which this larva operates on the apple-tree is by tying together the leaves with silken cords, forming a mass of considerable size, inside of which it lives gregariously, skeletonizing the leaves that it has thus appropriated, and filling them with its gunpowder-like excrement. It was so abundant in 1868, in some orchards, as nearly to strip many trees, especially in young orchards that were in an unthrifty condition. It is quite different from the Rascal Leaf-crumpler (*Phycita nebulo*), which lives all the time in a little black horn-like case; whereas this larva carries no house on its back. Moreover, the Leaf-crumpler is solitary in its habits; whereas this species live in communities of several dozen during their entire larval life. As to the moths produced from these two larvæ, they are as different from each other as a goat is from a sheep.

The only reliable way to prevent the depredations of such pests is to trap the moths in bottles, to catch the larvæ and kill them, and tear down their nests before they are half built. Birds will destroy a great many when the worms are very small.

The Army-worm (*Noctua unipuncta*).—This noxious insect is hatched from an egg deposited by the parent moth

Fig. 111.

Army-worm Moth.

(Fig. 111) at the base of perennial grass-stalks. The eggs hatch, in different localities, from May until the middle of July. Some entomologists affirm that these moths seek low meadows to deposit their eggs. The eggs are consequently deposited over a greater area of territory; and if the succeeding year prove wet, and favorable to the growth of the worms, we shall have the abnormal condition of their appearing on our higher and dryer lands, and of their marching from one field to another. For just as soon as the green grass is devoured in any particular field in which they may have hatched, these worms are forced, both from hunger and from their sensibility to the sun's rays, to leave the denuded field. When they have become nearly full-grown, and have stripped bare the fields in which they were born, and commence to march, they necessarily attract attention; for they are then exceedingly voracious, devouring more during the last three or four days of their worm-life than they had done during the whole of their previous existence. As soon as they are full-grown, they burrow into the earth, and, of course, are never seen again as worms.

Their increase and decrease is dependent on even more potent influences than those of a climatic nature. The worms are attacked by at least eight different parasites; and when we understand how persistent these last are, and how thoroughly they accomplish their murderous work, we cease to wonder at the almost total annihilation of the Army-worm the year following its appearance in such hosts. In the words of the late J. Kirkpatrick, "Their undue increase but combines the assaults of their enemies, and thus brings them within bounds again." We must also bear in mind that, besides these parasitic insects, there are some cannibal insects, such as the Fiery Ground-beetle, which come in for their share of this dainty food; while the worms, when hard pushed, will even devour each other. It is stated that Army-worm Moths do sometimes lay their eggs before harvest upon growing grain, sufficiently high from the ground for the egg to be carried off with the straw, which accounts for several well-authenticated instances of the Army-worm starting from stack-yards.

The Army-worm larva varies but little in appearance from the time it hatches to the time when it is full-grown. Some specimens are a shade darker than others, but the markings are generally uniform, as shown by Fig. 112, on the following page, which represents an Army-worm of full

Fig. 112.

An Army-worm.

size. The general color is dingy black, and it is striped longitudinally as follows: On the back a broad dusky stripe; then a narrow black line; then a narrow white line; then a yellowish stripe; then a narrow sub-obsolete white line; then a dusky stripe; then a narrow white line; then a yellowish stripe; then a sub-obsolete white line; belly obscure green.

Fish-brine for Exterminating the Aphides. — Brine and salt must be used around fruit-trees with extreme caution, lest the trees be killed before the insects are dispatched. We have frequently read of the successful use of weak fish-brine, sprinkled among the leaves and tender branches; but we have never used brine in any form, as we dare not, even when it is much diluted. Salt and brine will often kill trees and plants in a few days, without affecting a single insect in the least. It is a mistaken idea that salt spread around fruit-trees will repel or exterminate insects or worms of any kind, unless it is applied in sufficient quantities to destroy the last vestige of vegetation. Brine will doubtless destroy the aphides when sprinkled on the leaves of trees. But it must be much diluted, or the saline material will kill every leaf. Whenever brine is employed, the twigs and leaves should be sprinkled with water soon after the brine is applied. Brine or soap will perform its work on the tender lice in a few minutes, when fresh water should wash the leaves clean.

The Sulphur Remedy.—Some pomologists, both in America and in the Old World, have written favorable accounts of the use of sulphur on apple-trees for exterminating the aphides and some other depredators. The sulphur can be readily applied to small trees, by means of such a bellows as is employed to spread sulphur on grape-vines (Fig. 113). Such bellows may be obtained at most hardware stores.

Fig. 113.

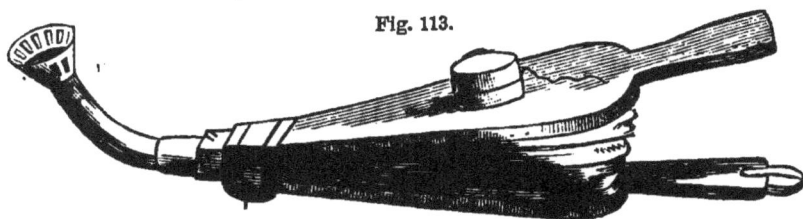

A sulphur-bellows.

The Red-humped prominent Caterpillar.—This formidable depredator—denominated scientifically *Notodonta concinna*—is rarely met with in many States; these caterpillars seem to have a preference for rose-bushes and pear-trees When full-grown, they are about one and a quarter to two inches long, having no sting, no irritating hairs or prickles, such as are found on a few of our rarer worms; and they will not even bite, however much we may irritate and torment them. Fig. 114 represents the full-grown Red-humped larva feeding on the edge of a leaf.

Fig. 114.

The Red-humped Caterpillar.

Fig. 115 represents the parent insect, or moth, with wings. Wherever this invader appears, the worms are extensively found, because the mother-moth deposits a very large number of eggs upon a single leaf. As these larvæ are gregarious throughout their entire existence, and do not

Fig. 115.

Parent moth of Red-humped Caterpillar.

scatter over the whole tree, as do many others that occur on our fruit-trees—some of which wander off from the very earliest stage in their larval life, and others, as for example the common Tent-caterpillar, only towards the latter part of their existence in the larval state—they can always be easily destroyed. The larva of the Red-humped will be readily recognized by its great beauty. The true way to exterminate these foes is to trap the moth in the bottles, as recommended (Fig. 51, p. 117), crush the chrysalis whenever it may be found, and catch the full-grown larvæ and kill them.

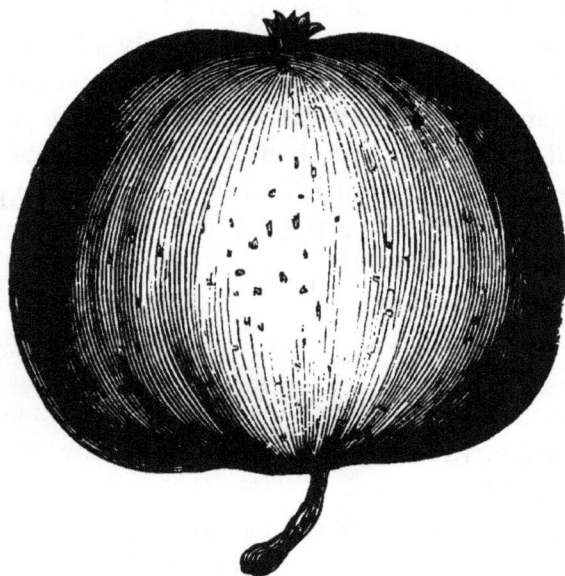

TALMAN'S SWEETING.

Synonyms.—Tallman's Sweeting, Tolman's Sweeting, and Brown's Golden Sweet. When fully ripe, those specimens which grow in the sunlight have a soft blush on one side; and, generally, a line runs from the stem to the calyx, which is set in a small basin slightly depressed. The flesh is quite white, rather firm, fine-grained, with a rich, sweet flavor. This is one of the very best winter sweet apples. The tree is a hardy, upright, and spreading grower. The apples are seldom formed like the engraving, with a swell on one side of the cavity. When the ground is rich, the apples grow much larger than represented by the cut. The Talman's Sweeting is a variety that may be added to every list of choice apples in every State of the Union, wherever apples can be raised with satisfactory success. This variety is a universal favorite with all people, and all kinds of domestic animals.

CHAPTER IX.

GATHERING AND MANAGEMENT OF APPLES.

With ladders, baskets, and the plucker's suit,
Pluck the fair King, fill baskets to the brim:
Relieve the bending boughs of luscious fruit:
Handle with care, and lay them in the bin.—EDWARDS.

APPLES, as well as apple-trees, possess vitality. There is life within the pulp, and life in the seeds. Let an apple be deprived of its vitality, and the once delicious pulp will become tasteless and unfit for eating. After an apple has been frozen and thawed, its life is gone, and with it every peculiarity of flavor and nutriment that made it pleasant to the taste, and a valuable and important article of food. It is the *life* of the apple, therefore, that we aim to preserve and to prolong through the changes of the weather that affect the fruit. Hence all the management of apples should be predicated on the assumption that careful handling and judicious management tend to prolong the life of the fruit. As violent treatment tends to destroy all animal life, so bruising and rough handling will hasten the decay of fruit. There is no life in a rotten apple, except within the seed.

These suggestions will enable us to understand something of the fundamental requirements for the successful preservation of choice apples during the unfavorable periods of alternate heat and cold. Hence, to maintain that condition of soundness which is essential to the value and excellence of apples, every specimen must be taken from the tree by the hand, and laid carefully in a basket. Winter apples and pears should be handled with as much care as one handles eggs. Apples *may*—and frequently do—fall to the ground without being injured. But a person has no assurance, if an apple falls to the ground, that it will not be so badly bruised as to hasten its decay in a few weeks. For this reason, hand-picking should be practised in preference to any other mode of gathering them, if to be kept for spring and summer use. When picked, lay those designed for long keeping carefully in the basket *with the hand*, instead of throwing them in. Winter apples should not be poured from one basket to another, any sooner than one would handle eggs by the rough process of pouring. Winter apples should never be handled in full bags, as any kind of fruit will be bruised more in a full bag than in any other receptacle. Transporting apples, even from the orchard to the fruit-room, in a common wagon, either before or after barrelling, is injurious; they should be moved on springs or on sleds. The least abrasion of the skin, or crushing of cells of the pulp containing the juice, induces fermentation and decomposition, and the consequent decay of the whole mass. Most orchardists who have many winter apples usually sustain heavy losses by the premature decay of the fruit, simply because they do not manage right with it. Fall apples may often be kept till winter, if plucked by hand and afterwards managed with proper care. The conditions for the largest and best preservation of apples and potatoes differ in several respects, and in one particular those conditions are in direct contrast. Apples require to be kept as *dry* as possible; potatoes to be kept moist.

When to pluck Winter Apples.—There is a variety of opinions as to the best period for gathering winter apples. Still there are certain facts as to time in which all do agree, one of which is, that all apples should be gathered before the weather is cold enough to freeze the ground. Apples, while on the trees, will not be injured by a light white frost. Winter apples will not freeze until the atmospheric temperature is from five to ten degrees below the freezing-point of water; and it is beneficial to keep them as cool as possible, even down to thirty degrees. Some persons have affirmed that apples inclosed in a water-tight cask may be left in a cold loft or garret all winter, without further care, and they will be sound in the spring, and perfectly fresh. But we have never dared to repose sufficient confidence in the doubtful experiment to try it. No matter where apples may be—if in a complete vacuum—if they freeze through, the frost will injure them. Another consideration of importance, touching the proper period for gathering winter apples, is this: the fruit should be plucked before a large proportion of it is blown off by the wind. If apples have ceased to grow, and are beginning to fall, the sooner they are plucked, the longer they will keep good. Some varieties should be gathered several weeks before the apples on the adjoining trees have attained their growth. Hence no person can indicate, on paper, the exact period when apples should be gathered. Winter.apples *can* be gathered too early; and there is danger of allowing them to remain on the trees too late in the growing season. When apples are ripe and the trees have ceased to grow, the fruit should be gathered at once, no matter what may be the day of the month. Winter apples should never be plucked when the fruit is wet, nor when it is chilled and almost frozen. Should an unexpectedly cold time occur, it will be better to let the fruit remain on the trees, than to shake it to the ground, unless it can be shaken *before* the apples are chilled. Every orchardist should make himself familiar with the qualities and characteristics of different varieties of winter apples, and pluck those first that ripen first. A little ob- servation on this point will enable one to determine with satisfactory certainty when is the proper time to gather different varieties.

Assorting Apples.—It is a common practice with many persons to mingle numerous varieties together. If such apples are alike in their season, there may be no objection to such management. But fall varieties and winter varieties should always be kept separate, so that a hundred bushels may not have to be overhauled for the purpose of picking out a few bushels which will soon decay and be lost, and hasten the decay of the sound apples, if they are not separated. Fall Pippins, for example, should be gathered before Spitzenbergs and the Baldwins, and should always be stored by themselves. Talman's Sweetings and other sweet apples should have each separate barrels or bins. Windfall should all be gathered before the main crop is plucked, and be stored in some convenient place for immediate use. A vast amount of economy may be exercised in the consumption of apples, by keeping the varieties and those of different qualities and conditions quite separate, when the fruit is gathered. When perishable varieties and windfall are mingled with fair winter fruit, servants, and sometimes the mistress of the family, will always choose the fairer fruit first; whereas, if the windfall, the bruised, and the perishable were selected, they would have better fruit for pies, dumplings, or for any other culinary purpose. In case one has a larger quantity of windfall and early fruit than can be readily disposed of before it decays, if it is not mingled with the best varieties, it can be fed to domestic animals, and thus be saved from immediate decay and loss.

Injury to Trees when gathering the Fruit. — We have alluded to this subject on a preceding page. The greatest care should be taken not to damage the trees when harvesting the fruit. The eye of the owner will be useful while this operation is being performed. Beginners and heedless helpers should be taught how to pluck an apple or pear when the stem separates with difficulty, so as not to break off the fruit-spurs or injure the buds, which are to produce the next year's crop. (See *Fruit-spurs*, p. 330.) There is a proper place for every stem to separate from the spur. A straight pull will often remove a piece of the twig several inches in length. When fruit is shaken off, the apples or pears often take long pieces of wood with them. Of course these have to be separated from the fruit; and it is far better to take a little pains and leave the wood on the tree. As the apple is about to be plucked, apply the thumb-nail to the stem at the proper place for separation, and break the stem across the nail. A great deal of damage is done to fruit-trees every year at the time of gathering the fruit. Large branches are trodden on and barked, small ones are broken, and, in the violent shaking of the trees, fruit-spurs are broken off. Selling fruit on the trees generally results in great injury to the orchard; for the buyer, in most cases, does not care how much damage is done to the trees, as his object is to gather the fruit in the most expeditious way. There is generally more damage done in gathering unripe fruit than in taking off that which is fully matured; for the former adheres so firmly to the spurs that some force is required to remove it.

Convenient Fruit-pluckers.—The illustration on the following page (Fig. 116) represents a convenient fruit-plucker, which is useful for plucking both pears and apples. For plucking valuable pears, it is almost indispensable, as it enables one to pluck all those good pears or apples on the ends of long limbs which could not be gathered except by shaking the trees. A plucker of this style may be made in a few minutes at a trifling expense. Bend a piece of wire in the form of the figure, as shown at *a*. No. 11 wire is

Fig. 116.

A cheap fruit-plucker.

about the right size. A piece of wire about thirty-two inches long will be required. Insert the ends in one end of a broom-handle, and then sew a little sack to the wire, large enough to hold about eight apples. The wire should be bent large enough to receive a man's hand into the sack. A blacksmith will fit one to a handle in four or five minutes; and, if a man has a pair of pliers, he can bend the wire to suit himself, after heating it in the stove. With one of this kind of fruit-pluckers, having a light handle eight or ten feet long, a man can pluck fruit rapidly while standing on the ground. When a person is in the tree-top, he should be provided with two or more pluckers, having handles of different lengths. Such fruit-pluckers are designed only for gathering such fruit as can not be reached with the hands. It will be understood that the fruit is pulled off by the narrow loop-end of the plucker.

Drying and seasoning Winter Apples. — As soon as the apples are plucked, they should be laid carefully in a cool, airy, and dry place, until there is danger that they will be injured by cold weather. A close apartment, which can not be ventilated sufficiently to carry away all moisture from the fruit, is an improper place for keeping winter apples during that period of autumn just before cold and freezing weather. If an apartment is so close that moisture will condense on the window-glass, it is a certain indication that the ventilation is quite insufficient for the requirements of the fruit. The best place for seasoning winter apples before cold weather is on a floor a few feet above the ground, and beneath a good roof, to carry off the storms, and with openings at the sides sufficient to admit a cool current of air both night and day. A building prepared like a tobacco-drying-house would be an excellent place for seasoning winter apples. Most cellars are quite too close and damp for winter apples until after the weather has become cold and freezing.

Apple-shelves in Dry Cellars.—When apples are kept in a cellar, one of the most satisfactory ways to keep apples is to make rows of shelves, or open bins, similar to Fig. 117, one above the other, with alleys between the rows, say three feet apart. Studs should be set up, to which narrow strips, three inches wide by one inch thick,

Fig. 117.

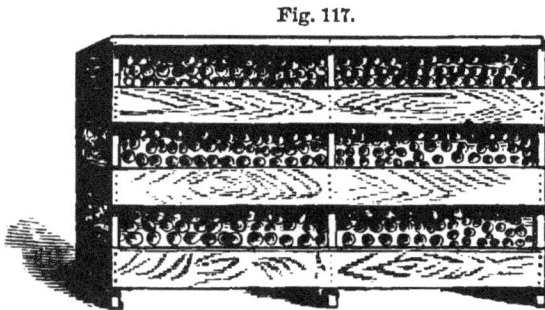

Convenient apple-shelves.

should be nailed. The lower shelf, or bin, should be at least six inches above the bottom of the cellar. The next about twenty inches above the first; and so on up to the joists. A person can walk on every side of such fruit-shelves, and can easily reach to the middle from either side. If rats and mice should gain access to such an apartment, they would find no refuge beneath bins beyond the reach of a cat. But a fruit-apartment should be made so tight that rats and mice can not enter, except through a door or windows. If they can avail themselves of no refuge, they will make a short stay in a cellar that is walled up so firmly that they can not find a hiding-place. Apples may be placed on each shelf, until the fruit is a foot or more deep. But it should be avoided, if possible, as it is in most cases unsafe; and with some varieties having thin, delicate skins,

13

most *certainly* so, as the fruit heats and specks in a short time. Give plenty of air, and all the circulation possible. An apple may be bruised in several places, and if it be kept in a close place it will begin to decay. On the contrary, if it is in an airy place, the bruises will soon begin to dry, so that a few dry days will dry and harden the bruises, and will keep the moisture from being absorbed from the wound. Windows at opposite sides or ends of such a fruit-cellar should be made, so that there may be no lack of ventilation at any season. There will be many days during the winter months when the windows should be thrown wide open day and night.

Stove in a Fruit-cellar.—We have assumed that the fruit-cellar, just alluded to, is beneath a dwelling-house, in which case it will be easy to provide a small stove for the fruit-room, to keep out the cold during severe weather. This practice will be found far more satisfactory than to attempt to bank up the outside walls, and close every door and window. A little fire in a small stove, placed near the outside door, will keep the temperature a few degrees above the freezing point, when without fire it would be almost impossible to prevent the fruit from freezing. The pipe from a small stove could pass up through the floor, and connect with the pipe or chimney in the first story. A far better way would be to let the chimney extend entirely to the bottom of the cellar. In cold and dreary localities, no one should attempt to keep fruit and vegetables in a cellar without a stove, to temper the atmosphere in very cold weather. In order to keep well, apples must not be exposed to severe cold, nor be kept too warm, neither be confined in a close apartment.

After cold weather is passed, winter apples will keep better in barrels, as many varieties, after warm weather has come on in the spring, will wilt up, become corky, and lose most of their flavor, making them comparatively worthless.

In a dry and well-ventilated cellar the air is constantly *renewed* and *kept dry*, thus carrying the damp vapor from the fruit as regularly as it escapes, by sweating or otherwise.

Storing Winter Apples in Barrels.—When apples are to be kept during the winter in barrels, after having been carefully hand-picked in baskets, the fruit should be laid on a floor, by hand, without pouring from the baskets, until they are twelve or eighteen inches deep, where the fruit should be left to dry and season three weeks, after which the apples should be carefully packed in clean dry barrels. The plan of drying and seasoning in the air before barrelling, prevailed generally some years ago, although nowadays it is mostly discontinued, and thought useless, as the process requires the exercise of too much care to comport with the fast notions of "Young America."

The following is a practice a pomologist recommends: Having picked your apples nicely, put them in the barrels without a leaf or straw or a spire of grass—head them up, and set them in a clean out-building where there is no offensive odor, and let them stand till the ground begins to freeze a little; then, in a clean grassy place, dig a ditch, square and straight, eighteen inches deep, and just wide enough to receive the barrels. In the bottom lay two fence rails close in the corners—pieces of rails will do. Now roll your barrels in, and they will be about six to eight inches above the surface lying end to end. Cover them with the dirt; but do not lay the sods on the barrels with the grass next to them. The covering may be ten inches thick or more; but you need not fear freezing, it won't hurt the apples. If it is a clay soil, you had better make your ditch on a slope, so that the water will run out at the lower end. When you want apples, dig out a barrel and put it in the cellar; and I warrant you will say the first you eat is best, for in ten days' time the exquisite aroma and taste will have departed, and the apples will be " only common apples." Many apple cultivators approve of the foregoing practice; but we have heard of so many apples failing to keep satisfactorily, that we do not indorse it as worthy of adoption.

Many people lay their winter apples carefully in barrels, keep them in a cool place until cold weather, then remove the barrels of fruit to a good cellar. To secure the apples against retaining moisture on the skin, let the opening of the windows, in dry states of the atmosphere only, be particularly attended to, as the circulation of dry air will soon absorb and carry off the natural moisture, while moist air will only add to it.

Removing the Rubbish.—In many orchards, the ground beneath most trees is strewn with pieces of brush, chunks of wood, and numerous small stones. To prevent damage to almost every apple that may fall to the ground, if it happens to strike a stone or stick, let all sticks, stones, and brush be removed from beneath every tree, so that the fruit will not be bruised when it falls from the trees. Stones and sticks should never be allowed to remain about fruit-trees; but the surface should be made smooth, and if it is covered with a coat of short grass, many good winter apples that are blown off the trees will not be injured when they fall. The fruit that falls first ought always to be kept by itself, so that it may be used late in the fall, or early in the winter, as such fruit will not usually keep so well as that which is plucked by hand. In some instances, straw is spread round about the tree for the fruit to fall on as the tree is shaken. The only objection to this practice is that, when fruit is shaken off the trees, much of it will be badly bruised by striking the branches of the tree-top, and also by falling on other specimens that may be on the ground.

Carrying Winter Apples in Baskets.—When apples are gathered within two or three minutes' walk of the fruit-room, the most convenient manner of conveying the fruit is to let two men carry two large basketfuls, as represented by the accompanying engraving (Fig. 118), as they will

be able to carry two large bas-kets of apples in this manner more easily than to lift them to their shoulders. It will be under-stood by the en-graving that a spar of timber, like a strong pitchfork handle, is put through the handles of each basket. If

Fig. 118.

Carrying two large baskets of winter apples.

the way is smooth, one person may take two large baskets on a wheelbarrow. Still another good way to convey ap-ples to the fruit-room is to set a dozen baskets on a large stone-boat. Winter apples should not be poured from one basket into another, nor be dumped into a wagon-box and jolted over a rough way, as all such rough handling will bruise them more or less, and thus hasten their decay.

Plucking from Tall Trees.—In many instances, apple-trees grow so tall that the limbs are not sufficiently strong to bear a small boy in the tree, nor on a ladder resting against it, unless the top is supported with guy-ropes. An orchard-ladder should have its lower ends shod with iron, in the form of a sharp wedge, to enter the ground readily, and to

Fig. 119.

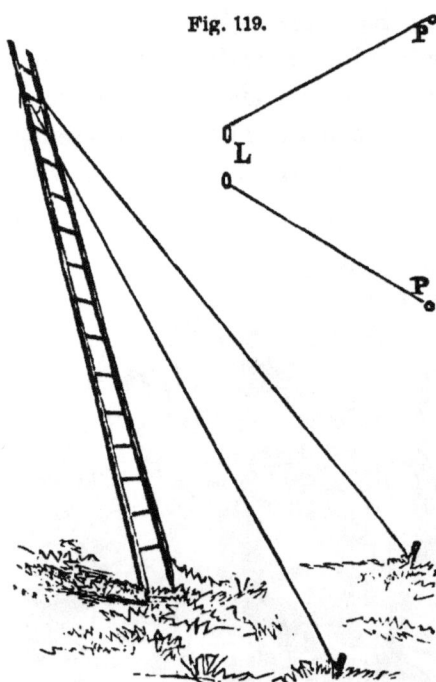

Manner of securing a long fruit-ladder by the
side of a tall tree, when it is desirable to
pluck the fruit from the long and slender
branches without injuring the buds and
twigs. L represents the top of the ladder,
and P P stakes driven into the ground, to
which cords from the upper end of the ladder are fastened.

hold the lower end when
putting it up and down.
Set the ladder nearly perpendicularly by the side
of a tree, and stay it with
two guy-ropes from the
top of the ladder, fastened to trees, or stakes, or
fences, as shown by Fig.
119. The ropes need not
be larger than a common
clothes-line. A man can
ascend to the very top of
a long ladder secured in
this way, and pluck half
a bushel or more of fruit
with entire safety.   A
large bag suspended on
one shoulder, and under
the arm on the opposite
side, is much more con-
venient than a basket, as
there is no danger of let-

ting the fruit drop, as with a basket; and both hands are
always free, whether the picker be in the tree or on a ladder.
If fruit be borne upon long, slender branches, by drawing
the ends inward or downward with a small hook, all the fruit
may be reached for several feet on each side.   The choicest
specimens of pears and apples often grow on the ends of
long, slender branches, which will not support a ladder, nor
a man while plucking the fruit. When long ladders are
leaned against the outside of trees, many of the small limbs
and fruit-buds are broken off. By securing a long ladder
as represented, fruit can be gathered by hand which could

not be reached in any other manner. Then, by providing a common grain-bag hung over one shoulder, as shown by Fig. 120, one can climb around in a tree-top with half a bushel of apples, pears, or other fruit, much more conveniently than if the fruit were in a basket.

Portable Staging. — The illustration given below (Fig. 121) represents a style of portable staging, which will be found exceedingly convenient when plucking any kind of

Fig. 120.

Plucking apples and putting them in a grain-bag hung over one shoulder.

Fig. 121.

Convenient portable fruit-ladder and staging.

fruit. It consists of two light ladders, about eight feet in length, having a strong board, say four feet long and fifteen inches wide, secured to the top of each ladder with small carriage-bolts. Nails would be liable to break when the staging is moved about. The rounds should not be more than a foot apart. If preferable, pieces of boards may be employed as *treads* instead of rounds. Let pieces be bolted in the form of the letter X, from the top of one leg to the bottom of another, to make *it stand firm.* Secure to one side strips for a hand-rail. When making such a device, procure small carriage-bolts at some hardware store, which will cost but little, and hold every part desirably firm. Every part should be made so light that one person will be able to move it from tree to tree. In case the ground is not level, let two extension legs be bolted to one side for the purpose of holding it level.

How to barrel Apples for Market.—The first apples that are put into a barrel should be laid in by hand, about as carefully as eggs are handled. It injures fruit, far more than most persons are wont to suppose, to pour it into a barrel. Even when a barrel is half-filled, the fruit will be injured by allowing one apple to drop on another. Fruit must be handled with extreme care. As soon as a barrel is full, let it be shaken thoroughly, press the head in with a screw or well-adjusted lever, like Fig. 122, with force enough

Fig. 122

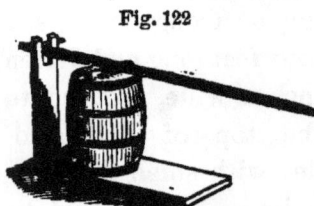

A convenient apple-packer.

to prevent the apples from moving when the barrel is transported, but not with force sufficient to break the skin or bruise the flesh of any of the fruit. If it were not for the necessity of packing the fruit so tight as to prevent its shaking when the barrel is moved about, it would be much better not to press the apples at all. The reckless manner in which apples are now pressed into their packages is one prolific cause of their premature decay. After the apples have been shaken down, allowance should be made for pressing the head down with the lever about half an inch. When apples are not pressed into the barrel so tightly that the fruit will not shake, almost every specimen will soon be badly damaged when transported on a wagon.

When to gather Apples.—Summer apples, and such fruit

as ripens early in autumn, will keep much longer if plucked before the specimens are dead ripe. The Sweet Bough and Early Harvest apples, for example, if gathered as soon as the fruit has really come to maturity, and laid carefully in a cool place, will often keep a month ; whereas, if left on the trees, they would become so dead ripe in a week that they would be of little value. And yet apples may be gathered too early. When fruit is gathered before it has come to maturity, it will wilt and become insipid, like a large proportion of the early apples that are sold in the New York markets. By a little observation, it will be noticed that well-matured apples, after being stored a while, look and feel quite oily when handled. They should be handled with considerable care after this process has taken place, so as to disturb this oily coating as little as possible, for this is Nature's covering to prevent decay; and what better material could she have supplied for such a purpose ? If apples are plucked before they have ceased to grow, this oily bloom will be wanting on the surface of the wilted fruit. Hence beginners should study to determine the true period for gathering every kind of fruit.

Fruit-preserving Houses.—In many large cities spacious fruit-preserving houses have been erected for the preservation of crude fruit until the succeeding summer. When one has but few apples, it would not pay to construct such a building. A fruit-preserving house is built very much like an ice-house. The side walls are made double, and the space between them is filled with sawdust or dry tan-bark. The doorway is double, having a door on each side of the jamb casings. The floor over the fruit-room is made of sheet metal, water-tight, and on it is placed ice two or three feet in depth, over which a deep layer of straw is spread. The ice maintains the temperature at such a uniform degree, that the more perishable sorts and varieties

13*

of fruit may be kept any desired length of time. The atmosphere in such a fruit-room is at about freezing-point, although the apples and other fruit are not frozen. As crude fruit is constantly emitting moisture, the fruit-room would soon become quite too damp, were it not for the *absorbent* material which is kept spread out on the floor in large quantities. Chloride of lime is employed for the absorbent, which has a great attraction for moisture, and a capacity for retaining a large quantity. As soon as this absorbent material is saturated with moisture, it is carried out into the air and dried thoroughly, and returned. By this means the fruit-room is kept cold and very dry. But when fruit is removed from such a preserving-house in the summer, it will soon decay, if it is not consumed immediately. If the moisture in such a fruit-room were not absorbed at once, the atmosphere would soon become like a dense aqueous and stagnant vapor, exceedingly unfavorable for the preservation of fruit.

Fried Apples for Human Food.—In those regions where potatoes are few and poor, on account of the rot, fried apples are an excellent substitute. They are quickly prepared for the table, which is often a consideration of no small importance. Wash them, cut them in two, take out the stem, core, and calyx, and, unpealed, put them into a tin pan with butter, or the gravy of baked pork, with some water, in proportion to the quantity to be fried; cover them with a lid, set them on the stove, stir them occasionally until they become soft, and be careful not to *burn* them. Romanites, which are often almost worthless, baked or raw, "disappear with good gusto when fried." We may truthfully pronounce despisable Penics good, when fried; but the Porters, Bellflowers, Talman's Sweeting, and a long list which we might name, when fried, are really a luxury. Most persons like apples in a raw state, or when they are baked or stewed; and it can not be denied that they furnish one of the most wholesome and agreeable kinds of diet. Although raw, stewed, or baked apples may be pronounced excellent and delicious, still there are many kinds, when fried, which are super-excellent; and they who seldom meet with a dish of fried apples often wonder why they are not always fried instead of baked. When apples are baked, they often burst open, and much of the best part flows out as juice, and is lost. But when they are fried, the whole is saved. In our own family we consume five or six barrels of apples before one barrel of Irish potatoes is gone. Many persons who do not care to eat more than one or two crude apples per day, will often eat six or eight when they are fried. Dr. Hall has stated, in his "Journal of Health," that, in order to derive a more decided medicinal effect, fruits should be largely eaten soon after rising in the morning, and about midway between breakfast and dinner.

Good ripe apples constitute a cooling diet. The philosophical reason for this is, the acid of the fruit stimulates the liver to greater activity in separating the bile from the blood, which is its proper work, the result of which is, the bowels become free, and the pores of the skin are open. Under such circumstances, fevers and want of appetite are impossible. The appetite frequently yearns for a pickle, when nothing else could be relished. This is often the case in the experience of most of us. It is the instinct of nature pointing to a cure. The want of a natural appetite is the result of the bile not being separated from the blood; and if not remedied, fever is inevitable, from the slightest grades to that of bilious, congestive, and yellow fever. But those persons who eat large quantities of crude or cooked apples are never troubled with constipation or biliousness. An incalculable amount of sickness and suffering would be prevented every year, if apples were employed to a greater extent on our tables, instead of such immense quantities of heavy animal food. Good beef, mutton, oysters, and roasted fowls make excellent living; yet, in many instances, a dish of fried apple will operate like magic in giving a healthful tone to the whole system of a high-living dyspeptic. By an arrangement of Providence as beautiful as it is benign, the fruits of the earth are ripening during the whole summer. From the delightful strawberry, on the opening of spring, to the luscious peach of the fall, there is a constant succession of superb aliments; made luxurious by that Power whose loving kindness is in all his works, in order to stimulate us to their highest cultivation, connecting with their use also the most health-giving influences. Liebig says, they prevent debility, strengthen digestion, correct the putrefactive tendencies of nitrogenous food, avert scurvy, and strengthen the power of productive labor. If eaten frequently at breakfast, with coarse bread and butter, without meat or flesh, apples have an admirable effect on the system, often removing constipation, correcting acidities, and cooling off febrile conditions more effectually than the most approved medicines.

Apples for Domestic Animals. — It will be perceived by the analysis on page 304 that sweet apples are of great value in feeding almost any kind of stock. Swine will fatten rapidly on them. Cows, if not over-fed with them at the start, and care is taken to cut or mash the apples, so that the animals can not get choked, will increase in milk and improve in condition. Apples are an excellent succulent food for horses in winter. Any varieties of sweet apples that bear abundantly will answer the purpose. Unfortunately, but little attention has been given to varieties for feeding domestic animals exclusively. Hardiness, thriftiness, and great bearing qualities are the main requisites. For early feeding, probably the High-top, or Summer Sweet of the West, may prove one of the best. There are several autumn sorts at least worthy of trial, among which are the following: the Coolies' Sweet is a fine grower, and a very productive New Jersey variety; the Jersey Sweet is also very productive, but the tree is less vigorous; the Munson Sweeting and Haskell Sweet are both productive, and of excellent quality, but not yet much tried west of New York. The Pumpkin Sweet may prove a good autumn sort for this purpose. The Sweet Pearmain is said to succeed well at the West, and may be valuable for late fall and winter feeding. The same remark will apply to the Sweet Romanite, a Western variety. The Wing Sweeting, although not large, is very productive at the East, and keeps well—if as much so at the West, it would be valuable. The Green Sweet is hardy and productive, and keeps into spring. The most profitable sweet apple that we have ever met with—both for market, for feeding stock, and for culinary purposes—is the Talman's Sweeting. (See Illustration, p. 282.) From a long experience in feeding sweet apples to milch-cows, horses, sheep, swine, and other stock, I *know* it will pay to cultivate the Talman's Sweeting to feed stock during the foddering season. If apples are fed in connection with meal, and only once per day, a bushel of apples will be of as much value as a bushel of oats. Yet apples can not be relied on as the proper feed for giving teams strength. When teams are standing idle for several days in succession, sweet apples, fed in connection with meal or grain, will keep animals in a better condition than nothing but grain and hay at every feeding. Most farmers have heretofore made an egregious mistake in judgment touching the value of sweet apples for feeding sheep during the foddering season, and for feeding store swine. A successful and intelligent farmer of Central New York recently wrote us as follows:

"I have been amused to hear farmers who have devoted years to their calling say that apples dry up their cows, and that they are not worth gathering for milch-cows. This year apples were too scarce to feed: but I think, from some experiments that I have made, that they are at least equal to carrots in weight for feeding stock, and especially milch-cows, as they always gain both in milk and flesh when fed on them. Apples and pumpkins should not be suffered to freeze, as that injures their feeding properties very much. I never saw lambs do better every way than when their feed was good hay and one bushel of apples per hundred sheep per day. Several who have fed apples to their sheep have refused to sell them for cider-making, and even fed out those of the best quality to dry, in preference to drying them for sale at the prices paid for dried fruit."

Chemical Analysis of Apples.—According to the analyses made by Professor Salisbury, "The apple is usually rich in phosphoric and sulphuric acids, and potash and soda. Hence we may infer that bone-dust, ashes, salt, and plaster would be likely to prove useful as portions of the manure applied to bearing trees, in addition to what is already contained in yard-manure. One striking difference in the composition of the apple and potato is the entire absence of starch in the apple, while in the potato starch constitutes nearly one-third of the solid part. The apple, according to this analysis, is rather superior to the potato in fat-producing qualities, which accords with the experience of some accurate farmers. The apple contains about twice as much of the compounds of nitrogen as the potato. The English Russet contains less water and more dry matter than most other sorts. This is doubtless the reason why this variety is so hard to freeze. The Talman's Sweeting contains more, the Greening still more, and the Kilham Hill most water of all, ranging in all these varieties from 79 to 86 per cent. A fresh potato contains about as much water as the Russet. The apple contains a small quantity of tannic and gallic acids, the proportion being greater in the Russets than in most other varieties. The astringency so striking in some kinds of apples, which is easily detected by the black color given to a knife or any iron substance when thrust into them, is owing to the presence of these acids. The ripe apple is rich in sugar and a body analogous to gum, called *dextrine*, which has the same composition as starch, although it differs from starch in being soluble in cold water, and not colored blue with iodine.

"Dextrine and sugar in the apple take the place of starch, dextrine, and sugar in the potato. Of the former, 100 lbs. of good fruit contain of dextrine, sugar, and extract 11.4 lbs.; the latter has, in the same amount of fresh tubers, 13.61 lbs. starch; dextrine, sugar, and extract, 68.5 lbs.; in the same quantity of dry potato there is of starch, dextrine, sugar, and extract, 68.02 lbs. The above proximate principles are the main bodies in the apple and potato which go to form fat. In the aggregate amount, then, of fat-producing products it will be seen that the apple and potato do not materially differ. It would be natural, however, to infer that 50 lbs. of dextrine and sugar would, if taken into the system, be more likely to make a greater quantity of fat in a given time, or at least to make the same amount in a shorter period, than an equal weight of starch, for this reason, that the two former bodies, although nearly the same in composition with the latter, are yet physically farther advanced in organization, and hence probably approximate nearer the constitution of fat. If this view be taken, then the apple, if of good quality, may be regarded equally, if not more rich in fat-producing products than the potato. One hundred lbs. of fresh apple contain of albumen 1.38 lbs.; the same amount of fresh potato has ¼ of a lb.; 100 lbs. of dry apple contain 9.37 lbs. of albumen, and an equal weight of dry tubers has 1¼ lbs.; 100 lbs. of fresh fruit contain of casein 0.16 of a lb., and an equal weight of fresh tubers 0.45 of a lb.; 100 lbs. of dry apples have 1 lb. of casein, and the same amount of dry potato contains 2¼ lbs. Hence it will be observed that 100 lbs. of fresh apple contain of albumen and casein 1.54 lbs., and the same quantity of fresh potato 0.7 of a lb.; 100 lbs. of dry fruit have of albumen and casein 9.37 lbs., and an equal amount of dry tubers contains 3.50 lbs.

"From the above it will readily be seen that in albumen the apple is richer than the potato, while in casein the reverse is the case; that the aggregate amount of albumen, casein, and gluten in good varieties of the apple is more than double that of the same bodies in the potato; hence the former may be regarded richer than the latter in those bodies which strictly go to nourish the system, or in other words, to form muscle, brain, nerve, and in short assist in building up and sustaining the organic part of all the tissues of the animal body.

"Dextrine and gum should not be confounded with each other. They differ very materially in many respects. The former possesses the property of being converted into grape-sugar by sulphuric acid and by diastase, while the latter does not. Dextrine belongs to the class of bodies which are susceptible of nourishing the animal body. All the starch taken as food is converted into dextrine before it is assimilated by the system. The acids of the stomach possess the property of converting starch into this body.

"In the fresh apple, 100 lbs. contain about 3.2 lbs. of fibre; 0.2 of a lb. of gluten, fat, and wax; 3.1 lbs. of dextrine; 8.3 lbs. of sugar and extract; 0.3 of a lb. of malic acid; 1.4 lbs. of albumen; 0.16 of a lb. of casein; and 82.66 lbs. of water. In the dry apple, 100 lbs. contain about 19 lbs. of fibre; 1.1 lbs. of gluten, fat, and wax; 18.7 lbs. of dextrine; 49.8 lbs. of sugar and extract; 2 lbs. of malic acid; 8.4 lbs. of albumen; and 1 lb. of casein. In the fresh potato, 100 lbs. contain about 9.7 lbs. of starch; 5.8 lbs. of fibre; 0.2 of a lb. of gluten; 0.08 of a lb. of fatty matter; ¼ of a lb. of albumen; 0.45 of a lb. of casein; 1.27 lbs. of dextrine; 2.64 lbs. of sugar and extract; and 79.7 lbs. of water. In the dry potato, 100 lbs. contain about 48.5 lbs. of starch; 29 lbs. of fibre. 1 lb. of gluten; 0.4 of a pound of fatty matter; 1.25 lbs. of albumen; 2.25 lbs. of casein; 6.32 lbs. of dextrine; and 13.2 lbs. of sugar and extract."

According to the analyses of Professor Salisbury, the apple has the advantage of containing a greater proportion of nitrogenous matter. Hence the value of apples, both as an article of human food and as food for all kinds of domestic animals. Professor Salisbury gives the inorganic and organic analyses of six different varieties of apples, viz.: Talman's Sweeting, Swaar, Roxbury Russet, Rhode Island Greening, and Kilham Hill. He also observes, that the analyses were calculated both with and without the carbonic acid. It was necessary that they should be calculated without it, in order to show the real composition of the organic matter of the fruit. The carbonic acid is formed during the combustion of the organic matter, and hence can not be regarded as a constituent part of the apple, except in very minute quantity. We extract the table showing the *mean* of those analyses, as follows:

INORGANIC OR ASH ANALYSIS.

	With Carbonic Acid.	Without Carbonic Acid.
Carbonic acid...............................	15.210	......
Silica......................................	1.362	1.637
Phosphate of iron...........................	1.336	1.593
Phosphoric acid	11.252	13.267
Lime.......................................	3.442	4.199
Magnesia...................................	1.400	1.669
Potash.....................................	31.810	37.610
Soda	20.810	24.799
Chlorine...................................	1.822	2.169
Sulphuric acid	6.062	7.229
Organic matter thrown down by nitrate of silver..	4.890	5.823
	99.396	100.000

PROXIMATE, OR ORGANIC ANALYSIS OF THE SAME VARIETIES.

	1000 Parts of fresh Apple.	1000 Parts of dried Apple.
Cellular fibre	32.03	190.879
Glutinous matter, with a little fat and wax.....	1.94	11.463
Dextrine	31.44	186.805
Sugar and extract	83.25	497.627
Malic acid	3.17	19.585
Albumen	13.79	83.720
Casein.....................................	1.64	9.921
Dry matter.................................	167.26	1000.000
Water......................................	826.64	......
Loss.......................................	6.10	......
	1000.000	......

Drying Apples.

How often, to dry, on foul cords are they strung,
In murky, low kitchens and out-houses hung;
As roosts for vile hornets, bugs, millers, and flies;
Then served at a banquet in dried-apple pies.—EDWARDS.

The practice adopted by a large proportion of those persons who prepare dried apples for market deserves the severest reprobation. The fruit is half-peeled, half-cored, and often not cored at all, cut in quarters or slices, and spread on the filthy roof of a building, or on dirty boards, where it is exposed to alternate sunshine and rain, until the repulsive-looking pieces, thickly dotted with fly-specks, are sufficiently dry to be stored in bins or barrels as an article of human food. The people who practise this odious system of drying fruit receive their merited reward in a price so low per pound that they have no encouragement to dry any more fruit. On the contrary, if they would peel their apples neatly, and dry them properly, so that the dry fruit would appear attractive as a desirable article of human food, instead of presenting a disgusting and repulsive appearance, they would receive an encouraging compensation for their labor. In one sense, the process of drying apples is like making butter. Let the labor be performed in a neat manner, and there will be no difficulty in obtaining a generous and a remunerative price. On the contrary, if neatness in milking, churning, and working the butter is neglected, the reward is a low price for the products of one's industry. The same is true in regard to drying apples.

The philosophy of drying apples, or any other fruit, consists in simply evaporating the water from the juice of the fruit, and curing the soluble portions of the fruit, so that they will not mould or decay. Hence, when fruit is dried by solar heat on scaffolds, the fundamental requirements are a rapid current of air, and the greatest amount of sunlight upon a limited extent of surface. To secure the first, the table or scaffold upon which the apple is to be spread should not be level, but inclined toward the south at an angle of ten or fifteen degrees, so as to present the surface more fully to the sun. The second point is gained by removing the skin and cutting the fruit into slices, the thinner, the more expeditious the result. The drying is facilitated by moving the pieces about several times during the day, so as to expose them more freely to the air and sun. They should be protected from the dew by a water-tight covering, as fruit will not dry during damp days and nights when exposed to the ordinary influences of the atmosphere. The process of drying must then be so rapid that no decay, nor even discoloration, shall take place until the operation is completed. Our climate is too precarious to think of drying fruit properly in the open air, even for the earliest varieties. Some artificial arrangement for the purpose must therefore be devised.

As soon as an apple is peeled, the sooner the water can be removed from the juice, the more delicious the dried fruit will appear. Ripe fruit must be dried rapidly, or it will begin to decay before the pieces are sufficiently dry to continue sweet during the entire year. Apples are frequently spread out on boards, and placed in an oven to hasten the drying process. The only objection to heating the fruit in an oven is a want of circulation. Heated damp air tends *to cook* the fruit rather than to dry it. If there could be a rapid current of warm *dry* air passing through the oven, so as to carry away the dampness from the fruit, a large oven would be an excellent place for drying fruit. When fruit is heated in a close oven, it is quite liable to be exposed to a degree of heat that will be injurious to it, so much so that it is often half-baked and afterwards dried. Hence such fruit is dark-colored, and deficient in that excellent aroma which dried fruit will possess when it has been dried in a current of rarefied or warm air. Rarefied air in motion possesses a wonderful capacity to absorb and convey away the moisture of fruit. When fruit is surrounded by warm air that is not in motion, decay will soon commence.

Fig. 123.

A convenient fruit-dryer.

A Cheap Fruit-drying Apparatus.—Figure 123 represents a cheap and convenient device for drying any kind of fruit, which may be employed in small families with the most satisfactory results. It consists of a strong box, two

by three feet square, or it may be only two feet square,
made of inch boards, without top or bottom. One side is
left open, as represented, to be closed by one or two doors
in front. The height may be two, four, or six feet. Cleats
are then screwed to two sides about two inches apart, to
support the drying-screens, which are made similar to the
sieves of a fanning - mill. Wire - cloth for making the
screens can be obtained in rolls of any desired length and
width at extensive hardware stores in large cities. If the
cupboard is made two by three feet square, wire-cloth two
feet wide can be procured and cut to fit the length of
the frames, and a piece nailed to each frame, as represented

Fig. 124.

A wire screen.

by Fig. 124. The frame-work of the screens
may be made of almost any kind of wood
dressed out smoothly, say one inch by one inch
and a quarter square, and the corners mitred,
glued, and nailed together, before the wire-cloth
is fastened to the frames. Coarse canvas may be
employed in lieu of wire-cloth. Yet wire will be found far
superior, as the circulation of warm air upward through the
wire meshes will be much more rapid than through canvas.
The screens should be about two inches apart. Then, in a
cupboard four feet high, there would be about twenty-four
screens, furnishing, in the aggregate, an area of scaffolding
equal to 144 square feet. When the dryer is to be used
for drying fruit of any sort, place it over any kind of stove,
a little above it, so that the heated air will rise directly upward through the meshes of the screens, and thus convey
rapidly away the moisture from the fruit. In the course
of a few hours, with a few dimes' worth of fuel, all the
fruit that could be spread on the screens would be dried in
the neatest manner. If the fruit at the bottom were to dry
more rapidly than that on the upper screens, they can be
changed in a moment. All steaming, stewing, and baking

is thus completely avoided. A current of dry, fresh, and warm air is constantly circulating through the box-cupboard; and so long as a particle of moisture remains, the dry air will absorb it.

An Apple-parer and Slicer. — Fig. 125 represents the latest and most approved apple-parer and slicer now in use. As fast as the apple is peeled, a peculiar-shaped knife cuts the entire pulp, except the core, in one long, spiral slice, somewhat like a very thick peeling placed on the edge. After an apple is peeled and sliced, with two cuts of a hand-knife every one is laid on the drying-

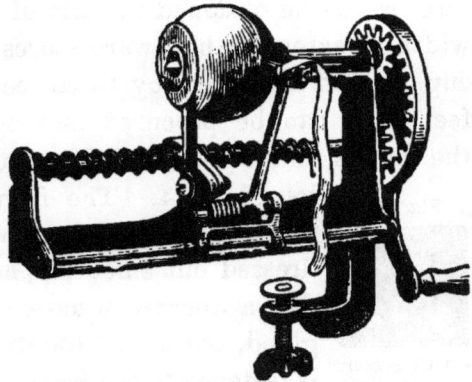

Fig. 125.

Paring, coring, and slicing machine.

screen, in twenty to forty neat and thin slices of a uniform size, for drying evenly. With such a machine one person can prepare several bushels per day for drying. They are very durable, and are not liable to get out of order. A small lad, with a little instruction, will soon learn to peel and slice apples quite as rapidly as an adult. One of the most important considerations is to keep the knives very sharp, so that they will cut easily through all bruised places. This style of parer was invented by D. H. Whittemore, of Worcester, Massachusetts; but the machines may be obtained of most dealers in hardware in large cities at about one dollar each.

The "Lightning Peeler." — Fig. 126, on the following page, is a fair representation of an improved "Lightning Peeler." The tines of the fork are secured to the journal in such a manner that they will clasp either a large or a small peach - pit. The knife moves automatically around the apple or peach, as represented by the dotted line. The main consideration in using a peeling - machine is to put each apple on the fork so that the outside

Fig. 126.

Combined peach and apple parer.

will revolve true, and to keep the knife as sharp as a razor, by whetting the edge on a fine-gritted whetstone.

Sweet-apple Molasses.—A great many well-to-do persons, who have large quantities of choice apples, have never tasted of apple molasses. If properly made, it is a superb luxury. To make a choice article, procure ripe, sound, sweet apples, reject all poor and half-rotten specimens— grind them, express the juice as directed for making superior cider, and boil the cider down to the desired consistency in a clean brass kettle, or in an iron kettle lined with mastic. Iron kettles will color the liquid. Maintain a steady fire, to prevent scorching it. The unfermented juice of any variety of sweet apples may be employed. While the sirup is being boiled, let all the scum be removed. As soon as it is of the desired consistency, put it in large bottles or kegs, and cork tight.

Sweet-apple Jelly.—One of the greatest luxuries of the apple-orchard consists of apple-jelly made of the juice of ripe, sweet apples, boiled down carefully as one boils maple-sap when making sugar, until it is sufficiently thick to form a jelly when it has cooled. If ripe, clean, and sound sweet apples be used, the jelly will have no taste of boiled cider, and will be of a beautiful amber color, and of a delicious taste. Some persons prefer the juice of tart apples to the juice of sweet fruit. They who have never tasted of such jelly will be surprised to see what a superb and delicate luxury can be made from the juice of apples. Such apple-jelly is not affected by exposure to the air, whether dry or moist, and it will neither sour, nor mould, nor dry up, nor absorb water. Such an article of course will bear transportation in barrels or other vessels to any part of the world.

When a person is going on a long journey, a few tin cans of such jelly can be carried conveniently; and the luscious food will always supply a great want. During hot weather, a spoonful mingled with water will make a beverage that kings and queens might covet. Some farmers prepare a number of gallons of such jelly expressly to mingle with drinking-water in hot weather. A few cents' worth of such jelly, mingled with water in a hot day, will enable a laborer to endure oppressive heat and fatiguing labor with very little inconvenience and suffering.

Sweet-apple Pies.—Pies made of sweet apples used in precisely the same way as pumpkins, omitting the ginger and adding a little lemon if liked, for seasoning, are better, to the writer's taste, than pumpkin-pie itself. Pare, cut, and stew the apples. If cooked in a covered deep earthen or other dish in the oven, they are better. Strain through a colander, adding a little milk, or cream, which is better. If there be no eggs to spare, stir in a handful of flour, or about a spoonful to a pie. Sweeten to taste with sugar; a mere trifle will be found sufficient. Bake thoroughly in a moderate oven.

How to make superior Cider.—In order to be able to express all the juice, every atom of an apple must be crushed.

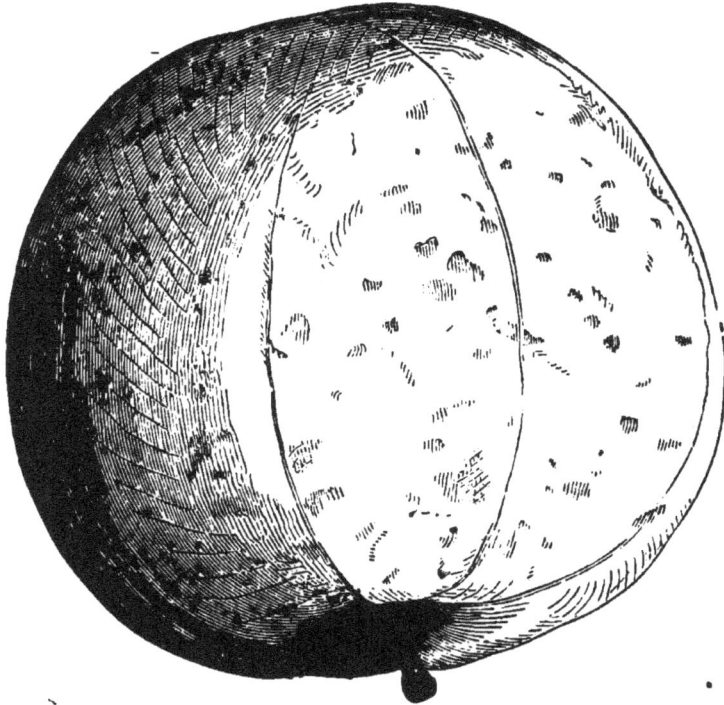

FALL PIPPIN.

Synonyms.—York Pippin, Pound Pippin, Episcopal, Golden Pippin, Cat Head, Philadelphia Pippin, and Pound Royal. This is a superb variety. The trees, when growing on good land, yield bountiful crops. This is really an autumn variety, while there are apples called Fall Pippins which are winter sorts. The fruit is usually large, roundish-*oblong*, somewhat angular, with broad ribs on its sides, terminating in an uneven crown, where it is nearly as broad as at the base. Calyx large, open, deeply sunk in a broad-angled, oblique, irregular basin. Stalk half an inch long, set in a rather small, even cavity. Skin smooth, yellowish-green on the shaded side, orange, tinged with brownish-red next the sun, and sprinkled with blackish dots; it is often covered with a bloom like that on a plum. Flesh yellowish-white, crisp, tender, with a sugary juice.

If sliced fine, the juice can not be forced out except by means of a powerful hydraulic press. When the apples are *squeezed*, as it is termed, without first being ground in a power-press, the cider is thin and light-colored, and of a very inferior quality; fit for nothing except for making alcoholic beverages. In order to make pleasant and aromatic cider, the apples must be reduced to a fine pulp destitute of lumps. After grinding or crushing, the pomace should be allowed to stand a few hours, according to the temperature of the weather, for the purpose of developing a rich color and a fine aromatic taste. In warm weather, ten hours is a sufficient length of time for the pomace to stand. As soon as little bubbles begin to appear on the surface, which indicate the first stages of fermentation, the juice should be expressed. When treated in this manner, the cider will possess a flavor which it never has when expressed as soon as the fruit is ground. Apple-seeds should never be crushed, as they will render the cider bitter. The apples should be well-ripened, but not in the least decayed. Every apple with the least speck of rot in it should be removed, if you wish a first-rate beverage. The decayed and inferior apples may be reserved for making vinegar. Perfect cleanliness should be observed in the grinding process. Clean dry straw should be used in forming the cheese. If the straw be musty, the flavor will be communicated to the juice. If water be added, it will make it hard and unpleasant to the taste. The casks, also, in which it is put for fermentation should be thoroughly cleansed, and finished off with a fumigation of brimstone. This is done by burning inside the barrel a few strips of canvas dipped in melted brimstone. The fumes will penetrate all the pores, and destroy the must and correct the sourness. After the fermentation is over, draw off into clean barrels, and clarify it. This can be done by mixing a quart of clean white sand with the whites of half a dozen eggs and a pint of mustard-seed, and pouring it into the barrel. It may stand in the barrel, or, if a nice article is wanted, it should be put into quart bottles and corked. When superb cider is made in the foregoing manner and secured in bottles, it is as palatable, and much more wholesome, than most of the wines of commerce. ¯In affections of the kidneys, it is an excellent remedy, and should have a place in every well-appointed cellar. It is a matter of some importance that what cider is made should be made in the best manner.

Cider and Wine Mills.—Fig. 127 represents a cheap ci-der-mill, designed particu-larly for those persons who desire to make only a few gallons of cider, or a few gallons of wine. The ap-ples are ground by hand, and the pomace falls di-rectly into the curb where the juice is to be ex-pressed. The grinding ap-paratus is all made of cast-iron, and every part is very durable. Such a mill will be found convenient when one desires to make only part of a barrel of cider in the latter part of summer, or in autumn, when it would not pay to make the same quantity with a large mill. The same press may be em-

Fig. 127.

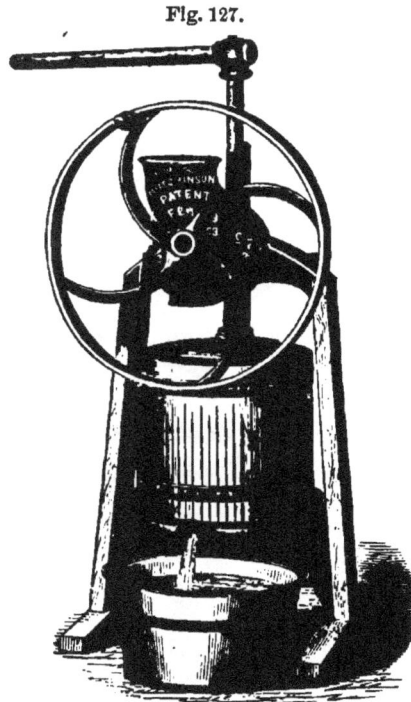

Hutchinson's hand cider and wine mill.

ployed for many other purposes. Grapes can be run through the grinder, or the juice expressed without grinding.

Making Cider-vinegar.—This liquid is a modification of *acetic acid.* There are many kinds of acetic acid sold as vinegar, which are so unlike cider-vinegar, that the vile stuff is no more fit to mingle with human food and drink than muriatic acid. Good cider-vinegar is a very useful condi-ment, and often an important luxury on the table. Let it

14

be understood, however, that we discard the hurtful acid that is made of numerous other substances besides the juice of apples, and sold as good vinegar. Cider-vinegar is made by exciting a second, or acetous, fermentation in cider. During this process, oxygen is absorbed from the atmosphere, carbonic acid is evolved, and the alcohol of the cider passes into acetic acid. In order to have cider-vinegar of the first quality, one must have good cider. The better the cider is, the better the vinegar will be. If vinegar be made of watered cider, it will be thin and weak, resembling watered cider. The accompanying illustration (Fig. 128) will

Fig. 128.

Apparatus for making cider-vinegar.

give the reader an excellent idea of the process of making good vinegar. A kind of cupboard is made of inch boards, about three and a half feet high by seven feet long. Inside of this box fit shelves about three and a half inches apart. On the upper side of these shelves gouge out channels running nearly from one end to the other, until the upper side is covered with zigzag grooves running from end to end. There should be cleats fastened to the under side of each shelf, to prevent it from warping; and the cleats should be put on with screws. The channel must be made slightly slanting, as in the illustration. The top shelf must slant so as to be about two inches lower than the other side, and the next shelf below it should slant about two inches in the opposite direction. By this arrangement, a long zigzag channel is made for the liquid to flow in. At its end, in the upper shelf, bore a hole through, so that the vinegar can drop to the next shelf, and traverse the channel. Thus it continues to flow from end to end, until it has reached the end of the channel in the lower shelf, when it falls into a receptacle. When commencing to make vinegar in this manner, place the maker in some small room, where you can have a fire and keep the temperature about ninety or ninety-five degrees Fahrenheit. Have a barrel, or tub, or hogshead, placed a little higher than the box, and near the end where the first channel commences, in the top shelf. In this barrel have a faucet, so that you can regulate the amount of cider which it is designed to have flow in the channel. The aim should be to keep a very small stream moving gently through the maker, affording every drop ample opportunity to absorb the desired amount of oxygen before the liquid reaches the end of the channel in the last shelf. A few gallons, or a half-barrel of good strong vinegar, should be run through first, so that the shelves will be well soured before letting other mixtures run through. It is a good idea to add one-third or one-fourth of good vinegar to any mixture of cider before allowing it to run through the maker. Open the faucet, so that a stream not larger than a straw shall fall into the channel of the top shelf. As it falls through the last hole into the barrel below the maker, the cider will have changed to strong and pure vinegar. When once started, the process must continue night and day, until the supply fails. In warm weather no fire will be required in the vinegar apartment, which should be well supplied with fresh air to facilitate oxidation. If the liquid is allowed to flow too rapidly, it will not have time to oxidize.

Buffalo Vinegar.—This kind of vinegar is made by mingling ten gallons of water with one gallon of molasses, and allowing it to ferment, after which one gallon of cheap whisky or two quarts of alcohol are added to every ten gallons. This liquid is then allowed to flow gently through the maker. This is the "Buffalo pure cider-vinegar," of which such large quantities have been sold. People buying it believe it to be made of pure cider; yet those in the trade know it does not contain one drop of cider. If a little more whisky be added, the vinegar will be stronger. By the addition of carbon in the form of molasses and sugar, the acid becomes more like cider-vinegar. Watered cider may be greatly improved for making vinegar by adding two gallons of molasses to a barrel of cider.

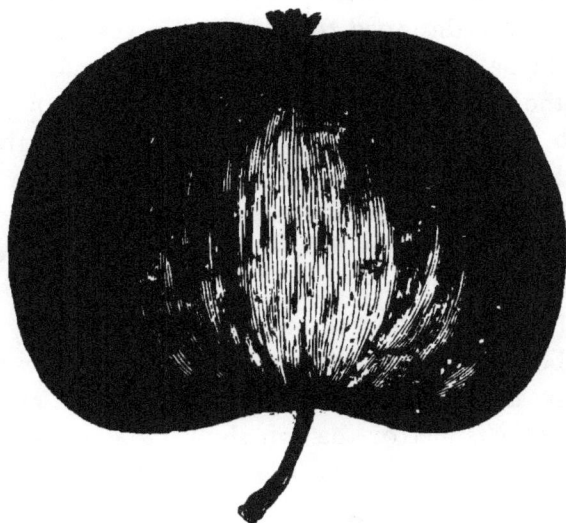

NEWTOWN SPITZENBERG.

Synonyms.—Spitzenburgh, Burlington, Kountz, Barrett's Spitzenburgh, Wine, Vandevere of New York, Ox-eye, Spiced Ox-eye, Joe Berry, and Matchless. The fruit of this old and valuable variety is usually of medium size, oblate, slightly conical, and, when not covered with stripes, splashes, and dots, the surface is of a fine yellow color, and is usually washed with red, is striped and splashed with deeper red, and richly shaded with carmine on the sunny side, covered with a light bloom, and sprinkled with grayish specks. The flesh is very rich and juicy, with a fine grain; crisp, tender, vinous flavor, scarcely subacid. Season, November to May.

CHAPTER X.

GENERAL PRINCIPLES OF POMOLOGY.

Tall branches stand waving their plumes to the sky,
To kiss the fleet summer clouds as they pass by;
And frequent they shower a roseate wreath
On innocent children while playing beneath.—EDWARDS.

AN abbreviated Pomological GLOSSARY is here present-
ed, embracing such words and phrases as are employed in
describing the apple, and in the propagation and management of apple-trees:

Acuminate, ending in a produced tapering point, having it curved towards one
edge of the leaf. (See *Leaf.*)

Albumen, nutritive organic matter, of great value as an article of food, and also
a valuable ingredient in manures. The white of eggs is almost pure albumen.
It constitutes the chief bulk of wheat, rye, and other monocotyledonous seeds.

Alburnum, the sap-wood of any tree, or the white portion between the inner bark
and the *duramen,* or heart-wood. Also the soft, semi-fluid substance—the *cambium*—between the *liber,* or inner bark, and the concentric circle of wood that was
formed the preceding year, after it has solidified.

Analogue, an organ or body resembling another organ or body, substituted for
or equivalent to it.

Analysis, comparing the various parts of a plant, tree, etc., with written de-
scriptions of a given specimen.

Annual, living or enduring but one year.

Annular, in the form of a ring.

Anther, that part of the stamen of a flower or apple-blossom which contains the
pollen. (See *Stamen.*)

Apetalous, destitute of petals; not having a corolla.

Apex, the crown, summit, or upper end, as the flower-end of an apple.

Apple. The scientific name of the common apple, including all known varieties,
is *Pyrus malus.* The pear belongs to the same genus as the apple, but is of a different species. Hence the pear is denominated *Pyrus communis.* Mrs. Lincoln, in her
"Botany," says the name of the apple is *Malus communis,* which is only one of
the mistakes of a great woman. The following is a brief botanical description of
the apple-tree: Stem, in open ground, ten to thirty feet high; in thickets, forty
to sixty feet high; branches rigid, crooked, and spreading; bark blackish and
rough; leaves two to four inches long, and two-thirds as wide as the length;
ovate, or oblong-ovate, serrate, acute, or short-acuminate, pubescent above,
tomentose beneath, petiolate; corymbs sub-umbellate; pedicels and calyx villose-
tomentose; pome (fruit) glohose; petioles one-half or one inch long; flowers expanding with the leaves, large, fragrant, and, when in full bloom, often clothing
the tree in a light, roseate hue. The blossoms of some trees, however, are nearly

white. (See *Blossoms.*)   There are at the present day over three thousand varieties of apples, and new varieties are being added to the list every year.   The timber of the apple-tree is very firm and heavy, and excellent for making planes and handles for tools; but it is not durable when exposed to the influences of wet and dry weather.   Botanically speaking,

Fig. 129.

Apex, crown, or blossom end.

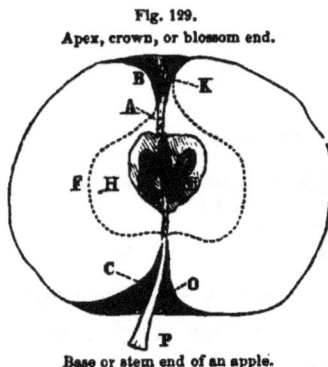

Base or stem end of an apple.

the *fruit* of the apple-tree is the seeds of the apple.   The names by which the principal parts of an apple are designated may be readily understood by the annexed diagram of a part of an apple, in which A represents the *axis* of an apple by a dotted line extending from the calyx, K, to the stem, O.   B represents the *basin* of the apple.   (See *Basin.*)   C shows the *cavity* of the fruit.   (See *Cavity.*)   K designates the *calyx;* F represents the edible portion of the apple, the *flesh* or *pulp,* and sometimes improperly called the *sarcocarp,* which is only applicable to stone fruit.   (See *Sarcocarp.*)   H shows the location of the *core,* which embraces the seeds, S, and the *hulls,* E, which separate the carpels, or cavities, in which the apple-seeds are formed—

the *hulls,* or membranaceous valves, which are tough, elastic shells, forming the inside walls of the carpels.   The number of seeds varies from one—or none at all—to every intermediate number between one and twenty.

Some apples may have produced even a larger number than twenty seeds.   A great many good apples have only a small core, and not a single perfect seed. The number of carpels, E, in an apple is five, which are all arranged around the *axis* of the apple.   All apple-seeds are *tunicated,* or covered with an *endocarp*—a number of concentric coats, which are sometimes so impervious to water that seeds will lie in the damp ground a year or more before a sufficient amount of moisture can find its way to the germ to promote germination.   The diagram herewith given represents a transverse section of an apple, showing the number of carpels, the seeds,

Fig. 130.

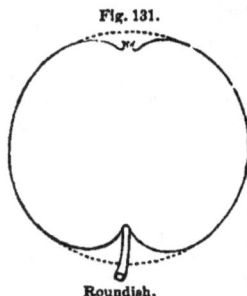

A section of an apple.

and the core.   It may be perceived, by a glance at this section, that when an apple is cut in transverse slices there will appear a perfect representation of the five petals of the apple-blossom.   The five dark-colored spots represent the carpels.   The black dot is the axis.   The illustrations herewith given will enable any intelligent person to describe any apple in an intelligible manner, by simply

Fig. 131.

Roundish.

comparing a specimen with the diagrams.   When the length and breadth of an apple are about equal, we say it is *round,* or *roundish,* like Fig. 131.   If the width is greater than the length from stem to calyx, the apple is of an *oblate* form, like Fig. 132, on the opposite page.   *Conical* is employed to describe an apple that is tapering from the base towards the apex, but which is not longer than the width, like Fig. 133. When the width is less than the length, like Fig. 134, the fruit is said to be *oblong.*   Kaighn's Spitzenberg is a variety of this form.   The Rhode Island Greening (p. 164) is described as *oblate-conical,* like Fig. 135. *Ribbed* apples are represented by Fig. 136.   Fruit of an *ovate* form is shown by Fig. 137.   The terms *red,*

Fig. 132.

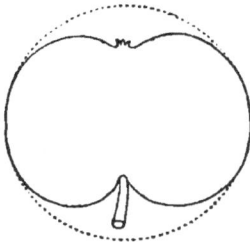

Oblate.

Fig. 133.

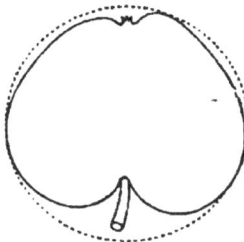

Conical.

Fig. 134.

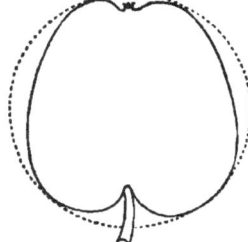

Oblong.

Fig. 135.

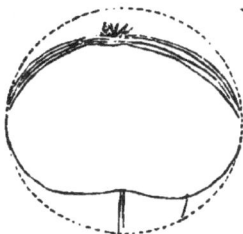

Oblate-conical.

Fig. 136.

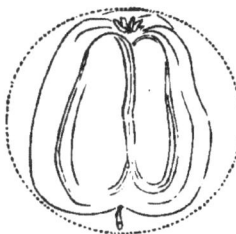

Ribbed.

Fig. 137.

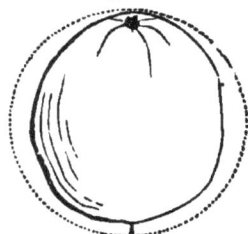

Ovate.

russet, striped, streaked, and several other adjectives, are employed when describing apples. .

Arbor, a tree, a perennial plant, the top, stem, and branches of which do not die annually, like the stem of an herb.

Arborescent, approaching the size and height of a tree.

Arboriculture, the art of propagating and cultivating trees and shrubs.

Armed, having thorns, spikes, or prickles.

Fig. 138.

Ascending form of branches.

Aromatic, having a spicy flavor or fragrance.

Ascending, rising from the ground in an oblique direction. When branches curve upward, like Fig. 138, they are said to be *ascending.*

Aspect, in the sense in which it is employed in pomology, signifies the slope of the land where an orchard is growing. When an orchard is planted on the north side of a hill, where the surface slopes to the north, we say the orchard has a northern *aspect.* If the surface slopes to the south or to the east, a *southern* or *eastern* aspect. Of course, when the land is level, the orchard will have *no* aspect.

Assimilate. The leaves of a tree change the crude sap into proper material for developing and building up every part, which is called *assimilating,* or digesting and concocting the plant-food.

Assurgent, rising in a curve from a declined base.

Axil, the angle between a leaf and a stem, or branch, on the upper side.

Axis, the central stem, or peduncle ; a central line extending from the base of an apple to the calyx. It sometimes signifies the elongated part of a petiole, upon which flowers appear.

Bark, the rind or covering of the woody parts of a tree. The bark of trees is composed of three distinct layers, of which the outermost is called the *epidermis,* the next, the *parenchyma*, and the innermost, or that in contact with the wood, the *cortical layer.* The epidermis is a thin, transparent, tough membrane; when rubbed off it is gradually reproduced, and in some trees it cracks and decays, and a fresh epidermis is formed, pushing outward the old; hence the reason why so many aged trees have a rough surface. The parenchyma is tender, succulent, and of a dark green. The cortical layer, or *liber,* consists of thin membranes encircling each other, and these seem to increase with the age of the plant. The liber, or inner bark, is known by its whiteness, great flexibility, toughness, and durability; the fibres in its structure are ligneous tubes. It is the part of the stem through which the juices descend, and the organ in which the generative sap, from whence all the other parts originate, is received from the leaves. The bark, in its interstices, is formed of cells, which are filled with juices of varying qualities; some, like that of the oak, remarkable for their astringency; others, like the cinnamon, abounding with an essential oil; others, as the Jesuits' bark, containing an alkali; some contain mucilaginous, and many resinous substances.

Base, the stem-end of an apple. (See *Apple.*)

Basin, the concavity, or depression of the end opposite the stem. The basin may be deep or shallow, broad or narrow. (See Fig. 129.)

Biennial, existing two years, or requiring two seasons to mature; or producing fruit once in two seasons.

Fig. 139.

Two apple-blossoms.

Bivalved, containing two valves.

Blotched, covered with different shades, which commence and disappear abruptly, without any regularity or order, over the surface.

Bloom, a delicate, powdery coating on the surface of fruit. It is of various colors, as per the fruit.

Blossom. An apple-blossom having only white petals is no more beautiful than a white rose; but an apple-blossom of many colors is a flower of exquisite beauty. By the aid of the accompanying diagrams, any person of only common-school education may soon gain a thorough botanical knowledge of the apple. The chief parts of an apple-blossom consist of the calyx, corolla, pistils, stamens, and peduncle. All these, taken collectively, constitute the flower, or blossom. Five of the central organs are *pistils.* (See *Pistils.*) The fine, thread-like organs are *stamens.* The five sections, or separate leaves of the corolla, are the *petals.* The green segments of leaves beneath the petals are denominated *sepals.* By cutting an apple-blossom in two equal parts, one can see, with the naked eye, at the bottom of the corolla, the miniature apple. By examining a perfect apple, we can see the five divisions of the dried-up calyx at the apex, or crown of the fruit. Every apple-blos-

Fig. 140.

A single apple-blossom.

som is a self-fertilizer. When the anthers on the ends of the stamens burst, some of the pollen from them will be scattered on the stigmas of the pistils, the surfaces of which are covered with a delicate adhesive liquid, which absorbs the pollen. Were it not for this process, no fruit would be perfected. Were the stamens destroyed before the anthers burst and scatter the pollen, every apple-tree and pear-tree would be destitute of fruit. The anthers do not burst until after the flowers are in full bloom.

Buds. There are five or more distinct names given to different kinds of buds; namely, *flower*-buds, *leaf*-buds, and *fruit*-buds, which contain both the flower and the fruit, the *axillary* buds, and *terminal* buds. Sometimes the flower-bud and the fruit-bud are identical, as in the apple and some other kinds of fruit. Every person who is permitted to have any thing to do with any kind of fruit-trees should be taught the difference between leaf-buds and fruit-buds; and such persons should have a lively understanding of the eminent practical importance of protecting the fruit-buds when the fruit is being gathered. The buds that are formed this year on apple-trees must produce the crop the next season. *Adventitious Buds* are frequently spoken of. These are new buds put forth by the tree near the stub-end of a branch that has been cut off. Such buds appear at the ends of roots that have been mutilated and cut off smoothly. All the buds of a branch do not expand every year, especially if the growing tree does not stand in rich soil. A portion of the buds often remain nearly dormant, so that if those on each side of them should be destroyed, those comparatively dormant will expand. When it is desirable to have buds developed early in the summer for the purpose of inoculation, let the terminal bud be pinched back as often as it starts. By this means the side buds will be fully developed in the former part of the growing season; whereas, were the branches permitted to increase in length, the side buds would scarcely be developed at the end of the year. Branch-

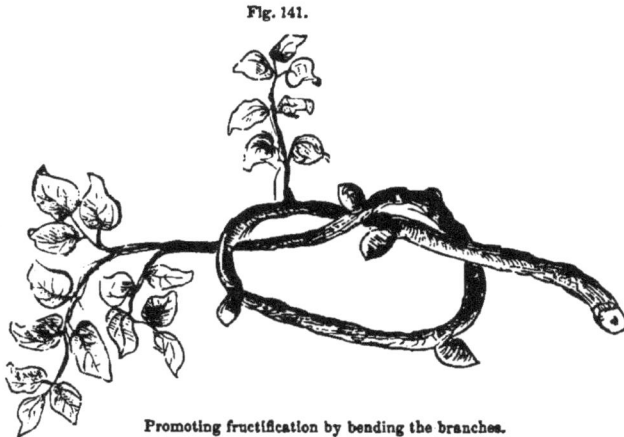

Fig. 141.

Promoting fructification by bending the branches.

es of some kinds of apple-trees are often inclined to grow long and slender, having very few small fruit-buds. By tying small ones in a large knot, like Fig. 141, fruit-buds will develop where the branch is bent. Other branches may be doubled backward towards the middle of the tree-top, and tied in any desirable position for the purpose of promoting fruitfulness.

Calyx, the most exterior integument in the very bottom of the basin of an apple. Some apples have a large, and others a small calyx. (See *Apple.*)

Cambium, or Cambium-layer, is the name that has hitherto been given to the descending sap which forms the mucilaginous annual deposit between the inner bark of the apple-tree and the outer ring of wood. So long as this material is liquid, semi-liquid, or plastic, it is called *cambium.* After it has solidified, even while it remains soft, it is spoken of as the *cambium-layer.* The cambium not

14*

only forms a new layer or ring of wood beneath the bark annually, but it renews the inner surface of the bark also. The words *cambium* and *alburnum* are frequently employed as if they were of the same signification. The two substances are about as nearly alike as ice and water.

Capsule, a dry, hollow seed-vessel, usually opening by regular valves and definite seams. It answers to the carpels of an apple.

Carnose, fleshy, more firm than pulpy.

Carpels. The carpels of the apple and pear consist of the seed-receptacles at the core, of which there are usually five, although botanists fix the number at two to five. They also assert that the pear (*Pyrus communis*) has two to five carpels.

Fig. 142.

A chrysalis.

When thinning the fruit of some of our own trees, we cut a few green pears into thin, transverse sections through the core, and found one pear having *six* perfect carpels, each containing two seeds.

Cavity, the depression of the centre of the stem-end of an apple, opposite the basin. The epithets *broad, narrow, deep, shallow, acute,* and *acuminate* are frequently employed in connection with the word cavity when describing apples.

Chrysalis, one of the forms of insects, like Fig. 142, from which they emerge with wings. It is more proper to speak of an insect in the *pupa* state, than to call it a chrysalis.

Cions, twigs, or *shoots,* signify portions of branches after one end has been fitted for the cleft in the stock. Then it also denotes the entire portion of a branch of the preceding season's growth. Cions are also called grafts, which can not properly be called cions. Fig. 143 represents a cion with one end sloped off, so that the name of the variety may be written on

Fig. 143.

A small cion.

it with a lead-pencil. Fig. 144 shows a bundle, as cions appear very soon after they have been cut from a tree. Such bundles should be buried in moist sand in a cellar, or in a sandy knoll, during cold weather. Fig. 145

Fig. 144.

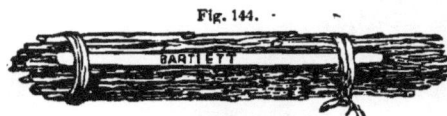

A bundle of cions labelled.

represents the manner of wrapping up a few cions with damp moss, in oiled paper, when they are to be sent by mail.

Cleft, a division between parts less than half-way to the base; also the crack that is formed in the end of a limb, or stock, for receiving a cion.

Fig. 145.

A bundle of cions to be sent by mail.

Collar, that part of the body or stem of an apple-tree near the surface of the ground. Some trees produce a broad and thick ridge at the surface of the ground, called the collar of the tree.

Concentric. Concentric circles, or layers, are the rings of wood, one of which is formed annually, around the body of every growing tree beneath the bark.

Conical, those apples that are more or less tapering, somewhat like a cone, from the base to the apex. (See Fig. 133.)

Cordate, heart-shaped, with the side-lobes rounded.

Corolla, the delicate inner covering of a flower between the calyx and stamens, usually colored, surrounding the parts of fructification, and composed of one or more flower-leaves, denominated petals.

Cortical, belonging to the bark of a tree.

Corymb, a mode of flowering; a sort of flat or convex flower-cluster.

Cotyledons, the seed-lobes, or fleshy portion of a kernel. A cotyledon involves and nourishes the embryo plant, and then perishes. Some seeds, like beans, pumpkin-seeds, apple-seeds, and many others, have two lobes, and are denominated *dicotyledonous,* because the two lobes of the seeds split apart during the process of germination; and each lobe appears on the end of the stem as a leaf. The lobes of the dicotyledonous seeds are flat on the inside, and convex on the outside. (See Fig. 1, p. 21.)

Crab-apple (Pyrus coronaria), is a wild apple-tree that produces small apples, very hard and exceedingly sour. The color of the fruit is usually a yellowish-white. The *Crab-apple of Siberia,* or the *Siberian Crab-apple,* is a hardy variety that may be cultivated in almost any locality, for the great beauty of the fruit, if for no other purpose. In some instances we have known this variety cultivated for market.

Fig. 146.

Crenate. This term is applied to leaves the edges of which are scalloped into rounded teeth.

Crown, the apex of an apple opposite the stem. (See *Apple.*)

Curculio, a general term for coleopterous insects, the larvæ of which destroy fruit. (See illustration, p. 238.)

Deciduous, falling at certain periods, like the leaves of apple-trees, which fall every autumn.

Disk, the face, or surface, of a leaf; also, the face, or central part of a head of compound flowers.

Diverging, applies to such branches of a tree as grow at about an angle of forty-five degrees from the stem. The Ribston Pippin-tree furnishes an example of diverging branches. Fig. 146 shows diverging branches.

Dotted, covered with dots that are distinct and separate from each other.

Diverging branches.

Doucin (erroneously *Doucain*), signifies a variety of the dwarf-apple. The Doucin stock forms an apple-tree larger than the Paradise stock, but not so large as the common standard apple-trees. Dwarf apple-trees are produced by grafting the common apple into Doucin stocks, or into Paradise stocks.

Drooping, when the limbs of a tree fall below a horizontal line.

Drupe, stone fruit, a fleshy, spongy pericarp, like a plum, inclosing a hard pit.

Embryo. The embryo of an apple-seed consists of the rudimentary plantlet, *caulicle, radicle,* or *plumule.* The plumule shoots upward, and the radicle downward. Fig. 147 is a section of a seed of a peony, in which is shown a small embryo. Fig. 148 is the same embryo detached and magnified. Fig. 149 is a section of a seed of barberry, showing the embryo in the middle of the albumen. Fig. 150 is the embryo separated from the albumen. Fig. 151 is a section of potato-seed showing the embryo coiled in the albumen.

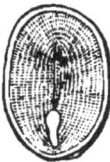
Fig. 147.

Fig. 149.

Fig. 151.

Fig. 153.

Fig. 148.

Fig. 150.

Fig 152.

Fig. 154.

ed in the albumen. Fig. 152 represents the embryo separated from the albumen. Fig. 153 shows a section of the Four-o'clock with the embryo coiled around the outside of the albumen. Fig. 154 is the embryo detached. The foregoing illustrations are copied, by permission, from "Gray's Botany."

Endocarp, the inner coating of the brown integument of an apple-seed. It also signifies the entire brown shell of an apple-seed. If we speak with scientific accuracy, the outer coat of the brown shell of an apple-seed signifies the *exocarp*, and the inner lining the *endocarp*. Endocarp signifies, also, the *putamen*, or the brown shell of a chestnut, and the stony shell of a peach, plum, and cherry.

Epicarp, the skin or peeling of an apple, which is a kind of epidermis.

Epidermis. The epidermis of a person's skin is the thin, insensible portion of the cuticle—the scarf-skin ; also the dry, shaggy part of the bark of an apple-tree. The epidermis covers every part of the tree that is exposed to the air, except the stigma of the blossom and the spongioles of the roots. The epidermis of the apple-leaf consists of the transparent skin, which seems like an elastic film of glass spread evenly over the surface. The very thin external covering of apples is the epidermis of the fruit.

Fig. 155.   *Epiphyte*, or *Epiphytes*, plants which grow on other plants, but which do not penetrate the living substance of the plant that supports them, to absorb any of the juices. Lichens or moss on the bark of an old apple-tree may properly be called *Epiphytes*, as the growing parts are supported by the dead bark of the tree.

Erect, when applied to shoots or branches of apple-trees, has reference to all those limbs shooting upward, between an angle of forty-five degrees and a perpendicular at the central stem of the tree. The branches of many varieties of pears, and some apples, grow erect. Fig. 155 shows erect branches.

Exogenous, growing on the outside, like an apple-tree, by adding annually one concentric layer of new wood beneath the bark.

Fecundation, impregnation ; the fertilization of a blossom by the union of the pollen with the stigma.

Filament, that part of the stamen of a flower which supports the anther.

Fissure, a slit, crack, or narrow opening.

Erect branches.   *Foliage*, the leaves of a tree, or the leaves and blossoms. Many reputedly intelligent pomologists say *foilage* for foliage.

Free, not adhering firmly to another part, as a peach is called a "free-stone" when it is not a "cling-stone," adhering firmly to the pit.

Free-stocks. Apple-seedlings are denominated *free*, or *free-stocks*, to distinguish them from the *dwarf*, or Doucin, and Paradise stocks, and from trees that have been produced by root-grafting. In other words, an apple-tree that sprang from a seed is a *free-stock*. If produced in any other way, it is not a free-stock.

Fructification, the flower and fruit, with their parts, consisting of the calyx or empalement, the corolla or petals, the stamens and the pistil, which belong to the flower, the pericarp and seeds, which pertain to the fruit, and the receptacle, or base, on which other parts are seated : the act of fructifying, or maturing fruit.

Fruit. In the common acceptation of the term, the word fruit means the edible portion of apples, pears—the fleshy part, or the *pericarp*. But, *botanically* speaking, fruit signifies the seeds of a turnip, the seeds of an apple, and the *kernels* containing the germ within the hard shell or pit of a peach or plum. The fruit of the potato consists of the small seeds in the balls which grow on the extremities of the vines. The fruit of the chestnut-tree is the chestnuts.

Genus, a group of species which agree with each other in the structure of essential characters of the flower and the fruit. Trees that agree in their flower and fruit, like the apple and the pear, are of the same genus, but of different species. The apple-tree, for example, and pear-tree both belong to the genus *Pyrus ;* but the apple belongs to the species *Malus*, and the pear to the species *Communis.*

Germ, the growing part of the bud : a point ; the miniature apple at the base of a blossom ; sometimes called an embryo ; the rudiment of an apple-tree in the seed of an apple, in an embryotic state. (See *Embryo.*)

Germination, the sprouting and first growth of a seed before the stem has reached the surface of the ground and formed leaves. After leaves have been formed, we say the plant *vegetates*, or grows.

Graft, or *Grafts.* A graft is often called a *cion*, and a cion a *graft*. But strictly speaking, a graft is the cion after it is inserted in the stock—the living, growing, partially, or fully-developed cion. (See *Cion*.)

Habitat, the natural abode of any animal, or that peculiar locality of plants or trees where they grow spontaneously.

Heeling-in, or *Laying-in-by-the-heels*, placing the roots of trees in a trench or hole, and covering them temporarily, while the stems of the trees are erect or inclined.

Hilum, the eye of a potato or point of an apple-seed. (See *Embryo*.)

Humus, vegetable mould formed by the thorough decomposition of vegetable matter in the soil. It is usually of a black color, and very fine, like dark-colored ashes.

Hybrid, when applied to animals, signifies a mule. Among plants and vegetables, a hybrid is the product of the union of two varieties of different species of trees, vines, or vegetables. A hybrid, also, is the product of the union of two individuals of different species; while a *cross* is the mixture of two varieties. Many intelligent persons employ the terms *hybridize* or *hybridizing* in the sense of *crossing*, and vice versa, which is incorrect. The product of the union of two varieties of strawberries is a cross—not a hybrid. The product of an apple and a pear-blossom would be a hybrid.

Insertion, the apex of the stalk or stem of an apple or of any other fruit.

Larva, an insect in the caterpillar state; the first stage after the egg in the metamorphoses of insects, preceding the pupa, or chrysalis, and perfect insect. The word *larva* is singular, and *larvæ* is plural. (See p. 238, *a*.)

Leaf, or *Leaves.* Leaves are usually the foliage of a tree. Sometimes the foliage embraces both leaves and blossoms. The accompanying illustration of an apple-leaf (Fig. 156) will furnish something of an idea of the wonderful wisdom there is in a leaf. We look at a leaf smilingly, and exclaim: "Well, there it is! It's nothing but a leaf! What can be said about a leaf?" Let us examine the various parts closely, and we shall doubtless meet with some practical suggestions which will enable tillers of the soil to produce better apples and more bountiful crops. The *blade* or *lamina* of a leaf embraces the entire leaf, except the stem. The upper end is the *apex* of the leaf; and B is the *base*. An apple-leaf is denominated a *simple* leaf, as there is but one on the petiole. The petioles of some

Fig. 156.

The principal parts of an apple-tree leaf.

trees, like the yellow locust, have many leaves on each petiole, and are hence called *compound* leaves. M represents the *mid-vein*, which is the principal prolongation of the petiole, P. The old name of the mid-vein is *mid-rib*. The primary branches, V, V, sent off from the mid-vein, are denominated *veinlets;* and the secondary branches, issuing from them are called *veinulets*. The leaves of the apple-tree are *serrate*, which see. In many apple-leaves the veinlets are opposite; and in others on the same tree, the veinlets will be alternate. The usual form of apple-leaves is *elliptical, ovate,* and frequently *oblong-ovate.* They are often *obicular.* The apex of an apple-leaf is often *acute,* or short *acuminate.* The upper side of apple-leaves is smooth, or glabrous. The under side is pubescent, or tomentose. The leaves of an apple-tree are both the *lungs* and the *organs of digestion.* No tree can exist any considerable length of time, during the growing season, without leaves. The mid-vein, veinlets, and veinulets, constitute the frame-work of the leaf, which is covered by the *parenchyma,* or cellular tissue. They are all conveyers of the vital fluid of a tree, just as the veins of the body of an animal convey the blood away

Fig. 157.

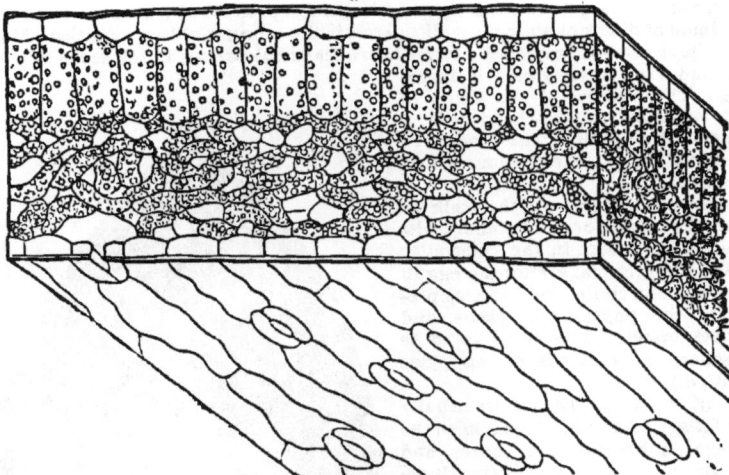

A highly-magnified section of a leaf.

from and back to the heart. The parenchyma of the leaves of an apple-tree is covered with a delicate varnish, which is impervious to water, as we may readily perceive by sprinkling water on the surface. When a tree is growing where there is a large amount of potash and silica in the soil, the leaves will be covered with a much thicker coat of this vegetable varnish than if potash and silica were scarce. Where these ingredients have nearly all been exhausted from the soil, the leaves of a tree will often be sickly, thin, and liable to be attacked by disease, simply because the roots can not find a supply of silica and potash to produce a heavy coat of varnish over the surface of both the leaves and fruit. Hence, when the spores of fungi come floating along in the air, the minute particles readily adhere to the leaves, and soon destroy them. Glass-makers employ silica, sand, and potash to manufacture glass. Apple-trees and other trees need a liberal supply of these materials to form a liquid similar to glass, to spread over the surface of leaves and fruit, to fortify every organ and tissue against the attacks of fungi. Let wood-ashes and sand be scattered round about apple-trees in great abundance, and the leaves will be of a dark-green color, tough like india-rubber, and the fruit will be free from rust and scabs. Fig. 157, given on the preceding page, is a fair representation of a section of a leaf magnified to show the air-chambers and the breathing-pores. Professor Gray states, in his "Botany," in connection with a similar figure, that in one square inch of the under side of an apple-leaf there are 24,000 breathing-pores.

Lobe, the division or segment of a petal or leaf; the free portion of a gamopetalous corolla; the cotyledons of a seed.

Longitudinal, or *Longitudinally*, from pole to pole, or from stem to calyx.

Malus, the scientific name of the species of trees to which the apple belongs.

Marbled, covered with wide, faint, waving, or irregular stripes.

Melting, becoming nearly a liquid and delicate pulp under a slight pressure, or when taken in the mouth, like a soft peach.

Mid-rib, the main, central nerve of a leaf, apparently the continuation of the petiole to the apex of the leaf. (See *Leaf*.)

Mottled. When an apple is covered with dots that appear to flow together.

Nerves, in leaves, rib-like fibres extending from the base towards the apex.

Oblate. An apple is of an oblate form when it is flattened, like Fig. 132, p. 319.

Oblate-conical. The Hawthornden, Rhode Island Greening, and some other apples, are said to be of an *oblate-conical* form (Fig. 135, p. 319).

Oblong, applying to such apples as Kaighn's Spitzenberg, having nearly parallel sides, and longer from stem to calyx than from side to side (Fig. 134, p. 319).

Oblong-conical, applying to an apple that is much longer from stem to calyx than from side to side, having the sides somewhat conical, like the Yellow Bellflower apple, p. 8.

Oblong-ovate, an apple more of the form of an egg than a cone; similar to the Black Gilliflower.

Obovate, inversely ovate, having the larger end at the apex of the fruit.

Obtuse, having rather blunt ends, or rounded off rather abruptly; not sharp.

Obtusely, in a rounded and blunt manner. The apple-leaf is *obtusely* toothed on the edge.

Ovary, the hollow portion at the base of the pistil, containing the ovules, or bodies destined to become seeds.

Ovate, somewhat like an egg. An apple of an ovate form is different from a conical form, in this respect, that the ovate is not tapered so much as an apple of a conical form. The Esopus Spitzenberg, Gilliflower, Porter, and some other varieties of apples, are ovate.

Ovules, the rudiments of future seeds contained in the ovary, or young fruit.

Paradise Apple, a small dwarf tree, scarcely larger than the currant-bush. When any of the varieties of the common apple are worked on the Paradise stock, a dwarf apple-tree is produced.

Parenchyma, the soft cellular tissue of leaves which covers the frame-work—the mid-vein, veinlets, and veinulets. (See *Leaf*.)

Pedicel, a partial peduncle; the ultimate branch, as in a compound inflorescence.

Peduncle, the stem of the apple-blossom, and also the stem of the apple itself. The peduncle, in some kinds of fruit, supports several pedicels, each of which bears a specimen of fruit. The stem of an apple has a base, and may be long or short, curved or straight, slender or thick, and it is sometimes knobby and fleshy. The stem characters are not very reliable.

Pentapetalous, having five petals, like the apple-blossom.

Perennial, living for a longer period than two or three years—like trees, grass, grape-vines, and shrubs.

Pericarp. The pericarp of an apple consists of all the parts outside of the seeds. That of a berry embraces the pulpy portion. In some fruits, the pericarp consists of the *epicarp, endocarp,* and the *sarcocarp.* The word *pericarp* is derived from two Greek words—*peri*, around, and *karpos*, the seed or fruit.

Petals, the delicate leaves of a flower or blossom. (See *Blossom*.)

Fig. 158.

Petiole, a foot-stalk or leaf-stem; not a penduncle, which is a fruit-stem.

Pistil, the central organ of a fertile flower, consisting usually of an ovary, *o;* the style, *s;* and the stigma, *a.* Some plants have only one pistil. The rose has numerous pistils. The embryo apple may be seen at *o*, in a perfect apple-blossom.

Pistillate, those flowers that have pistils, but no stamens.

Plumule, the young and tender stem of the future tree, when bearing two or more leaves. (See Fig. 3, p. 22.)

Pollen, the fine fertilizing powder contained in the anthers of a flower, without which a tree would produce no fruit, and plants no crops. We once cut off all the tassels of a hill of Indian corn, which grew alone in the yard of our city residence, before the tassel had grown above the leaves of the corn, and no corn grew on the cobs. If the stamens of apple-blossoms could be removed before the anthers burst, to allow the pollen to fall on the stigmas of the pistils, there would be no apples. The figures herewith given represent magnified views of pollen-grains, copied, by permission, from "Gray's Botany." Fig. 159 is a grain of the curious compound pollen of the pine. Fig. 160 (see below) is a pollen-grain from the flower of an Evening Primrose. Fig. 161 is a grain from the Enchanter's Nightshade. Fig. 162, a pollen-grain of the Kalmai flower. Fig. 163, a pollen-grain of the Succory. Fig. 164 represents the

Fig. 159.

A magnified pistil.

pollen of Wild Balsam Apple, which nearly resembles the pollen of the common apple when magnified. The pollen of certain plants, when magnified, possesses a more curious form than any of the accompanying fig-

Fig. 160.   Fig. 161.   Fig. 162.   Fig. 163.   Fig. 164.

ures. Professional botanists can often determine, by an examination of the form of the pollen grains, to what family the plant belonged, without seeing even the blossom that produced the pollen.

Pome, a fleshy, pulpy pericarp, containing one or more capsules, or carpels; as an apple or pear. The word *pome* is a term applied to apples and many other fruits.

Pomology, the science and art of propagating fruit-trees and cultivating fruit.

Primary, first in order of time or importance, as the primary roots of a plant, which are produced when the kernel first vegetates; opposed to the system of secondary roots of a plant or tree, which appear near the surface of the ground.

Pubescent, covered with very fine, soft hairs. The apple-leaf is pubescent above. (See *Tomentose*.)

Pulp, a soft, fleshy, juicy mass. It is often applied to the flesh of ripe fruit. Then it signifies crushed fruit or vegetables.

Pungent, sharp-pointed, or prickly at the apex; also acrid.

Pupa, an insect in the third, or next to the last state of existence, during which period it has not the power of locomotion; and when it takes no food. An insect in the chrysalis state. (See *Chrysalis*.)

Putamen, the hard shell of a walnut, or butternut, the brown shell of a chestnut, the brown shell of an apple-seed or pear-seed, the stone of a peach, plum, or cherry.

Pyramidal, tapering upward from the base, or bilge, to the top. When a hedge is sheared, or pruned of a pyramidal form, the sides taper upward from the base, or widest part, to a narrower top. But when a tree of any kind is pruned or sheared, the top above the bilge is more of a conical than of a pyramidal form.

Pyriform, largest at the apex, or crown of a fruit; shaped like a pear.

Pyrus, the scientific name of the *genus* of trees to which the apple and pear belong, while *malus* is the species.   Hence the name of the apple-tree—*Pyrus malus.* These are simply Latin names.

Quincunx. (See this explained on p. 74.)

Radicle, the slender, fibrous branches of roots.

Ramification, the profuse branching and subdivision of branches of roots or boughs.

Rhizoma, subterranean stems, or creeping roots.

Ribbed, having more or less ribs, or longitudinal ridges, running in parallel lines around fruit. (See Fig. 136, p. 319.)

Rigid, stiff, not pliable, inflexible.

Root, or *Roots.*   A root is an underground stem of a plant, shrub, or tree, which fixes itself in the ground, and serves to support the plant in an erect position. The roots which first appear from an apple-seed are denominated *seminal* or *primary* roots.   Those that are sent out from an apple-stem near the surface of the ground, in a horizontal position, are the *coronal*, or *secondary* roots.   A radicle is the first root of a kernel.   Then the radicles of an apple-tree are those fine hair-like roots which are sent out every growing season from the main roots to perform the office of *feeders* of the tree.   The radicles of an apple-tree are produced every year, and decay every autumn, like the leaves.   This fact, however, is controverted by most pomologists of the present age.

Sarcocarp, the fleshy, spongy, corky, or coriaceous part of the fruit, which is covered by the epicarp (from *sarx*, flesh, and *karpos*, fruit) ; the edible portion of the apple.

Seedling, the name given to an apple-tree that sprang from the seed, and which has never been grafted or inoculated ; an apple produced by a tree that has never been grafted.

Segments, the divisions, or separate portions of a circle, of a leaf, or cleft of the calyx of an apple.

Seminal, pertaining to seed ; rudimental ; issuing from the seed, as the seminal roots that start from a kernel of grain, or from an apple-seed.

Sepals, the leaflets beneath the petals of a flower.   They usually inclose the bud, and are of a green color.

Septum, or *Septa*, partitions that divide the interior of fruit.

Serratures, having the edge or margin notched somewhat like the teeth of a saw ; the sharp edges of the segments of leaves.   (See *Leaf.*)

Sheath, a tubular, membranaceous expansion of a plant that incloses a stem.

Shoots, the extremities of growing limbs or branches; particularly the part that grew the preceding year.

Shrub, a low dwarf tree or bush.   Correctly speaking, gooseberry-bushes and currant-bushes are shrubs, and not bushes.   Botanists have failed to designate the line of demarkation between trees and shrubs, so that a beginner can not always determine which is a tree and which a shrub.

Sinus, a bay ; a rounded cavity in the edge of a petal or leaf.

Spermoderm, the coarse shell, or skin, of an apple-seed, surrounding the kernel.   The three parts of the spermoderm consist of the external covering, called the *testa*, or *cuticle*, corresponding to the epicarp ; the cellular tissue, called the *mesosperm*, which corresponds to the sarcocarp of the apple ; and the thin inner skin, or *endosperm*, which is the same as the endocarp, or inside skin of the pericarp.

Splashed, having the stripes of all conceivable sizes, and much broken.

Spongioles, the delicate, soft, and sponge-like extremities of roots and radicles.

Spores, or *Sporules*, the analogues, or seminal equivalents of seeds in cryptogamous plants, which perform the functions of seeds.

Spray, the numerous small, fine twigs on the ends of branches.

Fig. 165.

Spreading, applied to the growth of branches of such trees as approach a horizontal direction, as shown by Fig. 165. The Rhode Island Greening apple-tree and many other varieties have spreading branches.

Sprouts are small, tall, and thrifty branches growing on the upper side of limbs, and sometimes from the body of an apple-tree.

Spurs (Fruit-spurs), short, stubbed branches, one to three inches long on the sides of the main branches, having one or more fruit-buds at the end.

Spreading branches.

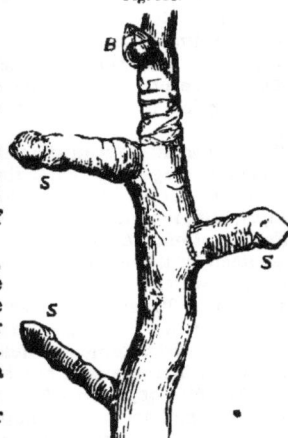

Fig. 166.

Fruit-spurs.

Fig. 166, at S S, represents fruit-spurs. At B is a fruit-bud.

Stained, an appearance of somewhat lighter shade than an apple that is colored or blotched.

Stamen, or *Stamens*, are the organs of a flower that prepare the pollen. They consist of the anther, *a*, and the filament, *b* (Fig. 167), situated between the petals and the pistils. The anther is a little case filled with dust, called pollen.

Fig. 167.

A stamen.

Stem, or *Peduncle*, in pomology, the cylindrical branch that supports the fruit. The stem sometimes signifies the body of a bush or tree.

Stigma, the summit of the style, or the portion of the pistil through which the pollen acts.

Stock, the stem or branch of either a young or old tree into which a cion or bud is inserted.

Straggling, applied to the growth of branches, like Fig. 168, which shoot out in almost every conceivable direction, like the branches of the Winter-nelis pear-tree. The forms of growth illustrated in this work have been made in accordance with those in Downing's "Fruit-trees of America," and J. J. Thomas's "Fruit Culturist."

Fig. 168.

Straggling form of growth.

Striped, when the stripes appear on an apple in alternate broad lines, like the Northern Spy, which is striped beautifully with red.

Sucker, or *Suckers*, off-sets, or shoots, either from the roots, stems, or branches of a tree.

Synonym, another name for the same thing. An apple, for example, often has several synonyms, or different names.

Tap-root, the large and strong root that extends directly downward into the earth, for the two-fold purpose of strengthening the position of the growing tree, and keeping it erect during storms and furious winds, and also to supply the leaves with moisture in hot and dry weather. It also exerts an important influence on the life and productiveness of apple-trees. (See Fig. 3, p. 22.)

Tenacious, adhesive, tough, sticky, or holding fast, or inclined to hold fast or retain a thing.

Terminal, situated at the very extremity or end.

Texture, referring to fruit as fine or coarse, tender, granular, or gritty, fibrous, tough, or hard.

Tissue, web or fabric, or the organic structure or composition of bodies.

Tomentose, covered with matted, woolly hairs; more matted than is expressed by the word *Pubescent*.

Transverse Section of an apple is made by cutting the specimen into two parts at nearly a right angle to the axis.

Truncate, or *Truncated*, having the extremity cut off or lopped, or one corner bevelled off.

Tunicated, covered with more or less membranes or coats, like the concentric coats of an onion.

Variety, or *Varieties*, properly signifies the difference between two individuals of the same species. In a loose way of talking, people speak of "different kinds" of apples, as the Fameuse, Jonathan, Talman's Sweeting, etc., which are *varieties* —*not* kinds. Kind signifies *genus*—*not* variety. Hence, in order to speak or write with scientific correctness, we should allude to apples and pears bearing different names as *varieties*.

Vertically, perpendicularly, or nearly so.

Villose, with soft, long, shaggy hairs.

Villose-tomentose is a compound term employed to describe the *calyx* and *pedicel* of the apple, signifying that such parts are covered with long, shaggy, soft, woolly, and matted hairs.

Worked on. When a cion or bud is inserted and continues to grow on a stock, the former is said to be "worked on the stock."

INDEX.

THE END.

www.ingramcontent.com/pod-product-compliance
Lightning Source LLC
Chambersburg PA
CBHW021459210326
41599CB00012B/1060